M000086708

Undergraduate Texts in Mathematics

Editors

S. Axler
F. W. Gehring
K. A. Ribet

Springer Science+Business Media, LLC

Springer Books on Elementary Mathematics by Serge Lang

MATH! Encounters with High School Students
1985, ISBN 96129-1

The Beauty of Doing Mathematics
1985, ISBN 96149-6

Geometry: A High School Course (with G. Murrow), **Second Edition**
1988, ISBN 96654-4

Basic Mathematics
1988, ISBN 96787-7

A First Course in Calculus, Fifth Edition
1986, ISBN 96201-8

Calculus of Several Variables, Third Edition
1987, ISBN 96405-3

Introduction to Linear Algebra, Second Edition
1986, ISBN 96205-0

Linear Algebra, Third Edition
1987, ISBN 96412-6

Undergraduate Algebra, Second Edition
1990, ISBN 97279-X

Undergraduate Analysis, Second Edition
1997, ISBN 94841-4

Complex Analysis, Fourth Edition
1999, ISBN 98592-1

Real and Functional Analysis, Third Edition
1993, ISBN 94001-4

Serge Lang

Introduction to Linear Algebra

Second Edition

With 66 Illustrations

 Springer

Serge Lang
Department of Mathematics
Yale University
New Haven, CT 06520
U.S.A.

Mathematics Subjects Classifications (2000):15-01

ISBN 978-1-4612-7002-7   ISBN 978-1-4612-1070-2 (eBook)
DOI 10.1007/978-1-4612-1070-2

Library of Congress Cataloging in Publication Data
Lang, Serge, 1927–
    Introduction to linear algebra.
    (Undergraduate texts in mathematics)
    Includes index.
    1. Algebras, Linear. I. Title. II. Series.
QA184.L37 1986   512'.5   85-14758

Printed on acid-free paper.

The first edition of this book was published by Addison-Wesley Publishing Company, Inc., in 1970.

9 8 7   SPIN 10977149

springeronline.com

Preface

This book is meant as a short text in linear algebra for a one-term course. Except for an occasional example or exercise the text is logically independent of calculus, and could be taught early. In practice, I expect it to be used mostly for students who have had two or three terms of calculus. The course could also be given simultaneously with, or immediately after, the first course in calculus.

I have included some examples concerning vector spaces of functions, but these could be omitted throughout without impairing the understanding of the rest of the book, for those who wish to concentrate exclusively on euclidean space. Furthermore, the reader who does not like $n = n$ can always assume that $n = 1, 2,$ or 3 and omit other interpretations. However, such a reader should note that using $n = n$ simplifies some formulas, say by making them shorter, and should get used to this as rapidly as possible. Furthermore, since one does want to cover both the case $n = 2$ and $n = 3$ at the very least, using n to denote either number avoids very tedious repetitions.

The first chapter is designed to serve several purposes. First, and most basically, it establishes the fundamental connection between linear algebra and geometric intuition. There are indeed two aspects (at least) to linear algebra: the formal manipulative aspect of computations with matrices, and the geometric interpretation. I do not wish to prejudice one in favor of the other, and I believe that grounding formal manipulations in geometric contexts gives a very valuable background for those who use linear algebra. Second, this first chapter gives immediately concrete examples, with coordinates, for linear combinations, perpendicularity, and other notions developed later in the book. In addition to the geometric context, discussion of these notions provides examples for subspaces, and also gives a fundamental interpretation for linear equations. Thus the first chapter gives a quick overview of many topics in the book. The content of the first chapter is also the most fundamental part of what is used in calculus courses concerning functions of several variables, which can do a lot of things without the more general matrices. If students have covered the material of Chapter I in another course, or if the instructor wishes to emphasize matrices right away, then the first chapter can be skipped, or can be used selectively for examples and motivation.

After this introductory chapter, we start with linear equations, matrices, and Gauss elimination. This chapter emphasizes computational aspects of linear algebra. Then we deal with vector spaces, linear maps and scalar products, and their relations to matrices. This mixes both the computational and theoretical aspects.

Determinants are treated much more briefly than in the first edition, and several proofs are omitted. Students interested in theory can refer to a more complete treatment in theoretical books on linear algebra.

I have included a chapter on eigenvalues and eigenvectors. This gives practice for notions studied previously, and leads into material which is used constantly in all parts of mathematics and its applications.

I am much indebted to Toby Orloff and Daniel Horn for their useful comments and corrections as they were teaching the course from a preliminary version of this book. I thank Allen Altman and Gimli Khazad for lists of corrections.

Contents

CHAPTER I

Vectors **1**

§1. Definition of Points in Space 1
§2. Located Vectors 9
§3. Scalar Product 12
§4. The Norm of a Vector 15
§5. Parametric Lines 30
§6. Planes 34

CHAPTER II

Matrices and Linear Equations **42**

§1. Matrices 43
§2. Multiplication of Matrices 47
§3. Homogeneous Linear Equations and Elimination 64
§4. Row Operations and Gauss Elimination 70
§5 Row Operations and Elementary Matrices 77
§6. Linear Combinations 85

CHAPTER III

Vector Spaces **88**

§1. Definitions 88
§2. Linear Combinations 93
§3. Convex Sets 99
§4. Linear Independence 104
§5. Dimension 110
§6. The Rank of a Matrix 115

CHAPTER IV
Linear Mappings **123**

§1. Mappings 123
§2. Linear Mappings 127
§3. The Kernel and Image of a Linear Map 136
§4. The Rank and Linear Equations Again 144
§5. The Matrix Associated with a Linear Map 150
Appendix: Change of Bases 154

CHAPTER V
Composition and Inverse Mappings **158**

§1. Composition of Linear Maps 158
§2. Inverses 164

CHAPTER VI
Scalar Products and Orthogonality **171**

§1. Scalar Products 171
§2. Orthogonal Bases 180
§3. Bilinear Maps and Matrices 190

CHAPTER VII
Determinants **195**

§1. Determinants of Order 2 195
§2. 3 × 3 and $n \times n$ Determinants 200
§3. The Rank of a Matrix and Subdeterminants 210
§4. Cramer's Rule 214
§5. Inverse of a Matrix 217
§6. Determinants as Area and Volume 221

CHAPTER VIII
Eigenvectors and Eigenvalues **233**

§1. Eigenvectors and Eigenvalues 233
§2. The Characteristic Polynomial 238
§3. Eigenvalues and Eigenvectors of Symmetric Matrices 250
§4. Diagonalization of a Symmetric Linear Map 255
Appendix. Complex Numbers 260

Answers to Exercises **265**

Index **291**

CHAPTER I

Vectors

The concept of a vector is basic for the study of functions of several variables. It provides geometric motivation for everything that follows. Hence the properties of vectors, both algebraic and geometric, will be discussed in full.

One significant feature of all the statements and proofs of this part is that they are neither easier nor harder to prove in 3-space than they are in 2-space.

I, §1. Definition of Points in Space

We know that a number can be used to represent a point on a line, once a unit length is selected.

A pair of numbers (i.e. a couple of numbers) (x, y) can be used to represent a point in the plane.

These can be pictured as follows:

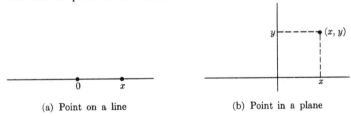

(a) Point on a line   (b) Point in a plane

Figure 1

We now observe that a triple of numbers (x, y, z) can be used to represent a point in space, that is 3-dimensional space, or 3-space. We simply introduce one more axis. Figure 2 illustrates this.

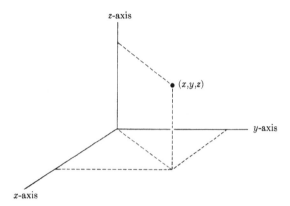

Figure 2

Instead of using x, y, z we could also use (x_1, x_2, x_3). The line could be called 1-space, and the plane could be called 2-space.

Thus we can say that a single number represents a point in 1-space. A couple represents a point in 2-space. A triple represents a point in 3-space.

Although we cannot draw a picture to go further, there is nothing to prevent us from considering a quadruple of numbers.

$$(x_1, x_2, x_3, x_4)$$

and decreeing that this is a point in 4-space. A quintuple would be a point in 5-space, then would come a sextuple, septuple, octuple, ...

We let ourselves be carried away and **define a point in n-space** to be an n-tuple of numbers

$$(x_1, x_2, \ldots, x_n),$$

if n is a positive integer. We shall denote such an n-tuple by a capital letter X, and try to keep small letters for numbers and capital letters for points. We call the numbers $x_1, \ldots, x_n$ the **coordinates** of the point X. For example, in 3-space, 2 is the first coordinate of the point $(2, 3, -4)$, and -4 is its third coordinate. We denote n-space by $\mathbf{R}^n$.

Most of our examples will take place when $n = 2$ or $n = 3$. Thus the reader may visualize either of these two cases throughout the book. However, three comments must be made.

First, we have to handle $n = 2$ and $n = 3$, so that in order to avoid a lot of repetitions, it is useful to have a notation which covers both these cases simultaneously, even if we often repeat the formulation of certain results separately for both cases.

Second, no theorem or formula is simpler by making the assumption that $n = 2$ or 3.

Third, the case $n = 4$ does occur in physics.

Example 1. One classical example of 3-space is of course the space we live in. After we have selected an origin and a coordinate system, we can describe the position of a point (body, particle, etc.) by 3 coordinates. Furthermore, as was known long ago, it is convenient to extend this space to a 4-dimensional space, with the fourth coordinate as time, the time origin being selected, say, as the birth of Christ—although this is purely arbitrary (it might be more convenient to select the birth of the solar system, or the birth of the earth as the origin, if we could determine these accurately). Then a point with negative time coordinate is a BC point, and a point with positive time coordinate is an AD point.

Don't get the idea that "time is *the* fourth dimension", however. The above 4-dimensional space is only one possible example. In economics, for instance, one uses a very different space, taking for coordinates, say, the number of dollars expended in an industry. For instance, we could deal with a 7-dimensional space with coordinates corresponding to the following industries:

1. Steel　　　　2. Auto　　　　3. Farm products　　　　4. Fish
5. Chemicals　　6. Clothing　　7. Transportation.

We agree that a megabuck per year is the unit of measurement. Then a point

$$(1{,}000,\ 800,\ 550,\ 300,\ 700,\ 200,\ 900)$$

in this 7-space would mean that the steel industry spent one billion dollars in the given year, and that the chemical industry spent 700 million dollars in that year.

The idea of regarding time as a fourth dimension is an old one. Already in the *Encyclopédie* of Diderot, dating back to the eighteenth century, d'Alembert writes in his article on "dimension":

Cette manière de considérer les quantités de plus de trois dimensions est aussi exacte que l'autre, car les lettres peuvent toujours être regardées comme représentant des nombres rationnels ou non. J'ai dit plus haut qu'il n'était pas possible de concevoir plus de trois dimensions. Un homme d'esprit de ma connaissance croit qu'on pourrait cependant regarder la durée comme une quatrième dimension, et que le produit temps par la solidité serait en quelque manière un produit de quatre dimensions; cette idée peut être contestée, mais elle a, ce me semble, quelque mérite, quand ce ne serait que celui de la nouveauté.

Encyclopédie, Vol. 4 (1754), p. 1010

Translated, this means:

> This way of considering quantities having more than three dimensions is just as right as the other, because algebraic letters can always be viewed as representing numbers, whether rational or not. I said above that it was not possible to conceive more than three dimensions. A clever gentleman with whom I am acquainted believes that nevertheless, one could view duration as a fourth dimension, and that the product time by solidity would be somehow a product of four dimensions. This idea may be challenged, but it has, it seems to me, some merit, were it only that of being new.

Observe how d'Alembert refers to a "clever gentleman" when he apparently means himself. He is being rather careful in proposing what must have been at the time a far out idea, which became more prevalent in the twentieth century.

D'Alembert also visualized clearly higher dimensional spaces as "products" of lower dimensional spaces. For instance, we can view 3-space as putting side by side the first two coordinates (x_1, x_2) and then the third x_3. Thus we write

$$\mathbf{R}^3 = \mathbf{R}^2 \times \mathbf{R}^1.$$

We use the product sign, which should not be confused with other "products", like the product of numbers. The word "product" is used in two contexts. Similarly, we can write

$$\mathbf{R}^4 = \mathbf{R}^3 \times \mathbf{R}^1.$$

There are other ways of expressing $\mathbf{R}^4$ as a product, namely

$$\mathbf{R}^4 = \mathbf{R}^2 \times \mathbf{R}^2.$$

This means that we view separately the first two coordinates (x_1, x_2) and the last two coordinates (x_3, x_4). We shall come back to such products later.

We shall now define how to add points. If A, B are two points, say in 3-space,

$$A = (a_1, a_2, a_3) \quad \text{and} \quad B = (b_1, b_2, b_3)$$

then we **define** $A + B$ to be the point whose coordinates are

$$A + B = (a_1 + b_1, a_2 + b_2, a_3 + b_3).$$

Example 2. In the plane, if $A = (1, 2)$ and $B = (-3, 5)$, then

$$A + B = (-2, 7).$$

In 3-space, if $A = (-1, \pi, 3)$ and $B = (\sqrt{2}, 7, -2)$, then

$$A + B = (\sqrt{2} - 1, \pi + 7, 1).$$

Using a neutral n to cover both the cases of 2-space and 3-space, the points would be written

$$A = (a_1, \ldots, a_n), \qquad B = (b_1, \ldots, b_n),$$

and we **define** $A + B$ to be the point whose coordinates are

$$(a_1 + b_1, \ldots, a_n + b_n).$$

We observe that the following rules are satisfied:

1. $(A + B) + C = A + (B + C)$.
2. $A + B = B + A$.
3. If we let

$$O = (0, 0, \ldots, 0)$$

be the point all of whose coordinates are 0, then

$$O + A = A + O = A$$

for all A.

4. Let $A = (a_1, \ldots, a_n)$ and let $-A = (-a_1, \ldots, -a_n)$. Then

$$A + (-A) = O.$$

All these properties are very simple, and are true because they are true for numbers, and addition of n-tuples is defined in terms of addition of their components, which are numbers.

Note. Do not confuse the number 0 and the n-tuple $(0, \ldots, 0)$. We usually denote this n-tuple by O, and also call it zero, because no difficulty can occur in practice.

We shall now interpret addition and multiplication by numbers geometrically in the plane (you can visualize simultaneously what happens in 3-space).

Example 3. Let $A = (2, 3)$ and $B = (-1, 1)$. Then

$$A + B = (1, 4).$$

The figure looks like a **parallelogram** (Fig. 3).

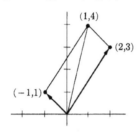

Figure 3

Example 4. Let $A = (3, 1)$ and $B = (1, 2)$. Then

$$A + B = (4, 3).$$

We see again that the geometric representation of our addition looks like a **parallelogram** (Fig. 4).

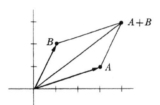

Figure 4

The reason why the figure looks like a **parallelogram** can be given in terms of plane geometry as follows. We obtain $B = (1, 2)$ by starting from the origin $O = (0, 0)$, and moving 1 unit to the right and 2 up. To get $A + B$, we start from A, and again move 1 unit to the right and 2 up. Thus the line segments between O and B, and between A and $A + B$ are the hypotenuses of right triangles whose corresponding legs are of the same length, and parallel. The above segments are therefore parallel and of the same length, as illustrated in Fig. 5.

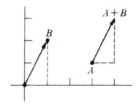

Figure 5

Example 5. If $A = (3, 1)$ again, then $-A = (-3, -1)$. If we plot this point, we see that $-A$ has opposite direction to A. We may view $-A$ as the reflection of A through the origin.

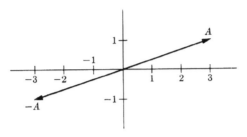

Figure 6

We shall now consider multiplication of A by a number. If c is any number, we **define** cA to be the point whose coordinates are

$$(ca_1, \ldots, ca_n).$$

Example 6. If $A = (2, -1, 5)$ and $c = 7$, then $cA = (14, -7, 35)$.

It is easy to verify the rules:

5. $c(A + B) = cA + cB$.
6. If c_1, c_2 are numbers, then

$$(c_1 + c_2)A = c_1 A + c_2 A \qquad \text{and} \qquad (c_1 c_2)A = c_1(c_2 A).$$

Also note that

$$(-1)A = -A.$$

What is the geometric representation of multiplication by a number?

Example 7. Let $A = (1, 2)$ and $c = 3$. Then

$$cA = (3, 6)$$

as in Fig. 7(a).

Multiplication by 3 amounts to stretching A by 3. Similarly, $\frac{1}{2}A$ amounts to stretching A by $\frac{1}{2}$, i.e. shrinking A to half its size. In general, if t is a number, $t > 0$, we interpret tA as a point in the same direction as A from the origin, but t times the distance. In fact, we define A and

B to have the **same direction** if there exists a number $c > 0$ such that $A = cB$. We emphasize that this means A and B have the same direction **with respect to the origin**. For simplicity of language, we omit the words "with respect to the origin".

Mulitiplication by a negative number reverses the direction. Thus $-3A$ would be represented as in Fig. 7(b).

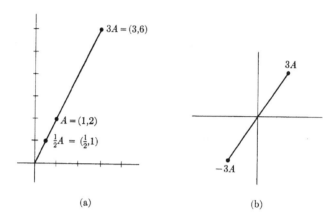

(a)   (b)

Figure 7

We define A, B (neither of which is zero) to have **opposite directions** if there is a number $c < 0$ such that $cA = B$. Thus when $B = -A$, then A, B have opposite direction.

Exercises I, §1

Find $A + B$, $A - B$, $3A$, $-2B$ in each of the following cases. Draw the points of Exercises 1 and 2 on a sheet of graph paper.

1. $A = (2, -1)$, $B = (-1, 1)$ 2. $A = (-1, 3)$, $B = (0, 4)$

3. $A = (2, -1, 5)$, $B = (-1, 1, 1)$ 4. $A = (-1, -2, 3)$, $B = (-1, 3, -4)$

5. $A = (\pi, 3, -1)$, $B = (2\pi, -3, 7)$ 6. $A = (15, -2, 4)$, $B = (\pi, 3, -1)$

7. Let $A = (1, 2)$ and $B = (3, 1)$. Draw $A + B$, $A + 2B$, $A + 3B$, $A - B$, $A - 2B$, $A - 3B$ on a sheet of graph paper.

8. Let A, B be as in Exercise 1. Draw the points $A + 2B$, $A + 3B$, $A - 2B$, $A - 3B$, $A + \frac{1}{2}B$ on a sheet of graph paper.

9. Let A and B be as drawn in Fig. 8. Draw the point $A - B$.

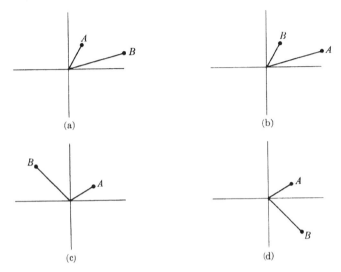

Figure 8

I, §2. Located Vectors

We define a **located vector** to be an ordered pair of points which we write $\overrightarrow{AB}$. (This is *not* a product.) We visualize this as an arrow between A and B. We call A the **beginning point** and B the **end point** of the located vector (Fig. 9).

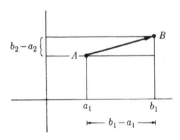

Figure 9

We observe that in the plane,

$$b_1 = a_1 + (b_1 - a_1).$$

Similarly,

$$b_2 = a_2 + (b_2 - a_2).$$

This means that

$$B = A + (B - A)$$

Let $\overrightarrow{AB}$ and $\overrightarrow{CD}$ be two located vectors. We shall say that they are **equivalent** if $B - A = D - C$. Every located vector $\overrightarrow{AB}$ is equivalent to one whose beginning point is the origin, because $\overrightarrow{AB}$ is equivalent to $\overrightarrow{O(B - A)}$. Clearly this is the only located vector whose beginning point is the origin and which is equivalent to $\overrightarrow{AB}$. If you visualize the parallelogram law in the plane, then it is clear that equivalence of two located vectors can be interpreted geometrically by saying that the lengths of the line segments determined by the pair of points are equal, and that the "directions" in which they point are the same.

In the next figures, we have drawn the located vectors $\overrightarrow{O(B - A)}$, $\overrightarrow{AB}$, and $\overrightarrow{O(A - B)}$, $\overrightarrow{BA}$.

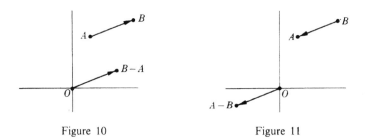

Figure 10   Figure 11

Example 1. Let $P = (1, -1, 3)$ and $Q = (2, 4, 1)$. Then $\overrightarrow{PQ}$ is equivalent to $\overrightarrow{OC}$, where $C = Q - P = (1, 5, -2)$. If

$$A = (4, -2, 5) \quad \text{and} \quad B = (5, 3, 3),$$

then $\overrightarrow{PQ}$ is equivalent to $\overrightarrow{AB}$ because

$$Q - P = B - A = (1, 5, -2).$$

Given a located vector $\overrightarrow{OC}$ whose beginning point is the origin, we shall say that it is **located at the origin**. Given any located vector $\overrightarrow{AB}$, we shall say that it is **located at** A.

A located vector at the origin is entirely determined by its end point. In view of this, we shall call an n-tuple either a point or a **vector**, depending on the interpretation which we have in mind.

Two located vectors $\overrightarrow{AB}$ and $\overrightarrow{PQ}$ are said to be **parallel** if there is a number $c \neq 0$ such that $B - A = c(Q - P)$. They are said to have the

same direction if there is a number $c > 0$ such that $B - A = c(Q - P)$, and have **opposite direction** if there is a number $c < 0$ such that

$$B - A = c(Q - P).$$

In the next pictures, we illustrate parallel located vectors.

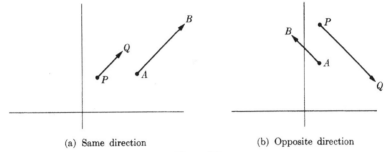

(a) Same direction   (b) Opposite direction

Figure 12

Example 2. Let

$$P = (3, 7) \quad \text{and} \quad Q = (-4, 2).$$

Let

$$A = (5, 1) \quad \text{and} \quad B = (-16, -14).$$

Then

$$Q - P = (-7, -5) \quad \text{and} \quad B - A = (-21, -15).$$

Hence $\overrightarrow{PQ}$ is parallel to $\overrightarrow{AB}$, because $B - A = 3(Q - P)$. Since $3 > 0$, we even see that $\overrightarrow{PQ}$ and $\overrightarrow{AB}$ have the same direction.

In a similar manner, any definition made concerning n-tuples can be carried over to located vectors. For instance, in the next section, we shall define what it means for n-tuples to be perpendicular.

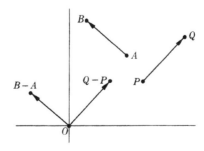

Figure 13

Then we can say that two located vectors $\overrightarrow{AB}$ and $\overrightarrow{PQ}$ are **perpendicular** if $B - A$ is perpendicular to $Q - P$. In Fig. 13, we have drawn a picture of such vectors in the plane.

Exercises I, §2

In each case, determine which located vectors $\overrightarrow{PQ}$ and $\overrightarrow{AB}$ are equivalent.

1. $P = (1, -1)$, $Q = (4, 3)$, $A = (-1, 5)$, $B = (5, 2)$.

2. $P = (1, 4)$, $Q = (-3, 5)$, $A = (5, 7)$, $B = (1, 8)$.

3. $P = (1, -1, 5)$, $Q = (-2, 3, -4)$, $A = (3, 1, 1)$, $B = (0, 5, 10)$.

4. $P = (2, 3, -4)$, $Q = (-1, 3, 5)$, $A = (-2, 3, -1)$, $B = (-5, 3, 8)$.

In each case, determine which located vectors $\overrightarrow{PQ}$ and $\overrightarrow{AB}$ are parallel.

5. $P = (1, -1)$, $Q = (4, 3)$, $A = (-1, 5)$, $B = (7, 1)$.

6. $P = (1, 4)$, $Q = (-3, 5)$, $A = (5, 7)$, $B = (9, 6)$.

7. $P = (1, -1, 5)$, $Q = (-2, 3, -4)$, $A = (3, 1, 1)$, $B = (-3, 9, -17)$.

8. $P = (2, 3, -4)$, $Q = (-1, 3, 5)$, $A = (-2, 3, -1)$, $B = (-11, 3, -28)$.

9. Draw the located vectors of Exercises 1, 2, 5, and 6 on a sheet of paper to illustrate these exercises. Also draw the located vectors $\overrightarrow{QP}$ and $\overrightarrow{BA}$. Draw the points $Q - P$, $B - A$, $P - Q$, and $A - B$.

I, §3. Scalar Product

It is understood that throughout a discussion we select vectors always in the same n-dimensional space. You may think of the cases $n = 2$ and $n = 3$ only.

In 2-space, let $A = (a_1, a_2)$ and $B = (b_1, b_2)$. We define their **scalar product** to be

$$A \cdot B = a_1 b_1 + a_2 b_2.$$

In 3-space, let $A = (a_1, a_2, a_3)$ and $B = (b_1, b_2, b_3)$. We define their **scalar product** to be

$$A \cdot B = a_1 b_1 + a_2 b_2 + a_3 b_3.$$

In n-space, covering both cases with one notation, let $A = (a_1, \ldots, a_n)$ and $B = (b_1, \ldots, b_n)$ be two vectors. We define their **scalar** or **dot product** $A \cdot B$ to be

$$a_1 b_1 + \cdots + a_n b_n.$$

This product is a **number**. For instance, if

$$A = (1, 3, -2) \quad \text{and} \quad B = (-1, 4, -3),$$

then

$$A \cdot B = -1 + 12 + 6 = 17.$$

For the moment, we do not give a geometric interpretation to this scalar product. We shall do this later. We derive first some important properties. The basic ones are:

SP 1. *We have* $A \cdot B = B \cdot A$.

SP 2. *If* A, B, C *are three vectors, then*

$$A \cdot (B + C) = A \cdot B + A \cdot C = (B + C) \cdot A.$$

SP 3. *If* x *is a number, then*

$$(xA) \cdot B = x(A \cdot B) \quad \text{and} \quad A \cdot (xB) = x(A \cdot B).$$

SP 4. *If* $A = O$ *is the zero vector, then* $A \cdot A = 0$, *and otherwise*

$$A \cdot A > 0.$$

We shall now prove these properties.
Concerning the first, we have

$$a_1 b_1 + \cdots + a_n b_n = b_1 a_1 + \cdots + b_n a_n,$$

because for any two numbers a, b, we have $ab = ba$. This proves the first property.
For **SP 2**, let $C = (c_1, \ldots, c_n)$. Then

$$B + C = (b_1 + c_1, \ldots, b_n + c_n)$$

and

$$A \cdot (B + C) = a_1(b_1 + c_1) + \cdots + a_n(b_n + c_n)$$
$$= a_1 b_1 + a_1 c_1 + \cdots + a_n b_n + a_n c_n.$$

Reordering the terms yields

$$a_1 b_1 + \cdots + a_n b_n + a_1 c_1 + \cdots + a_n c_n.$$

which is none other than $A \cdot B + A \cdot C$. This proves what we wanted. We leave property **SP 3** as an exercise.

Finally, for **SP 4**, we observe that if one coordinate a_i of A is not equal to 0, then there is a term $a_i^2 \neq 0$ and $a_i^2 > 0$ in the scalar product

$$A \cdot A = a_1^2 + \cdots + a_n^2.$$

Since every term is ≥ 0, it follows that the sum is > 0, as was to be shown.

In much of the work which we shall do concerning vectors, we shall use only the ordinary properties of addition, multiplication by numbers, and the four properties of the scalar product. We shall give a formal discussion of these later. For the moment, observe that there are other objects with which you are familiar and which can be added, subtracted, and multiplied by numbers, for instance the continuous functions on an interval $[a, b]$ (cf. Example 2 of Chapter VI, §1).

Instead of writing $A \cdot A$ for the scalar product of a vector with itself, it will be convenient to write also A^2. (This is the only instance when we allow ourselves such a notation. Thus A^3 has no meaning.) As an exercise, verify the following identities:

$$(A + B)^2 = A^2 + 2A \cdot B + B^2,$$

$$(A - B)^2 = A^2 - 2A \cdot B + B^2.$$

A dot product $A \cdot B$ may very well be equal to 0 without either A or B being the zero vector. For instance, let

$$A = (1, 2, 3) \qquad \text{and} \qquad B = (2, 1, -\tfrac{4}{3}).$$

Then

$$A \cdot B = 0$$

We define two vectors A, B to be **perpendicular** (or as we shall also say, **orthogonal**), if $A \cdot B = 0$. For the moment, it is not clear that in the plane, this definition coincides with our intuitive geometric notion of perpendicularity. We shall convince you that it does in the next section. Here we merely note an example. Say in $\mathbf{R}^3$, let

$$E_1 = (1, 0, 0), \qquad E_2 = (0, 1, 0), \qquad E_3 = (0, 0, 1)$$

be the three unit vectors, as shown on the diagram (Fig. 14).

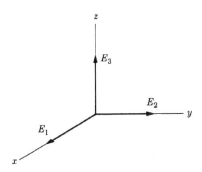

Figure 14

Then we see that $E_1 \cdot E_2 = 0$, and similarly $E_i \cdot E_j = 0$ if $i \neq j$. And these vectors look perpendicular. If $A = (a_1, a_2, a_3)$, then we observe that the i-th component of A, namely

$$a_i = A \cdot E_i$$

is the dot product of A with the i-th unit vector. We see that A is perpendicular to E_i (according to our definition of perpendicularity with the dot product) if and only if its i-th component is equal to 0.

Exercises I, §3

1. Find $A \cdot A$ for each of the following n-tuples.
   (a) $A = (2, -1)$, $B = (-1, 1)$ (b) $A = (-1, 3)$, $B = (0, 4)$
   (c) $A = (2, -1, 5)$, $B = (-1, 1, 1)$ (d) $A = (-1, -2, 3)$, $B = (-1, 3, -4)$
   (e) $A = (\pi, 3, -1)$, $B = (2\pi, -3, 7)$ (f) $A = (15, -2, 4)$, $B = (\pi, 3, -1)$

2. Find $A \cdot B$ for each of the above n-tuples.

3. Using only the four properties of the scalar product, verify in detail the identities given in the text for $(A + B)^2$ and $(A - B)^2$.

4. Which of the following pairs of vectors are perpendicular?
   (a) $(1, -1, 1)$ and $(2, 1, 5)$ (b) $(1, -1, 1)$ and $(2, 3, 1)$
   (c) $(-5, 2, 7)$ and $(3, -1, 2)$ (d) $(\pi, 2, 1)$ and $(2, -\pi, 0)$

5. Let A be a vector perpendicular to every vector X. Show that $A = 0$.

I, §4. The Norm of a Vector

We define the **norm** of a vector A, and denote by $\|A\|$, the number

$$\|A\| = \sqrt{A \cdot A}.$$

Since $A \cdot A \geq 0$, we can take the square root. The norm is also some-times called the **magnitude** of A.

When $n = 2$ and $A = (a, b)$, then

$$\|A\| = \sqrt{a^2 + b^2},$$

as in the following picture (Fig. 15).

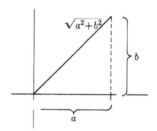

Figure 15

Example 1. If $A = (1, 2)$, then

$$\|A\| = \sqrt{1 + 4} = \sqrt{5}.$$

When $n = 3$ and $A = (a_1, a_2, a_3)$, then

$$\|A\| = \sqrt{a_1^2 + a_2^2 + a_3^2}.$$

Example 2. If $A = (-1, 2, 3)$, then

$$\|A\| = \sqrt{1 + 4 + 9} = \sqrt{14}.$$

If $n = 3$, then the picture looks like Fig. 16, with $A = (x, y, z)$.

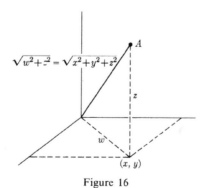

Figure 16

If we first look at the two components (x, y), then the length of the segment between $(0, 0)$ and (x, y) is equal to $w = \sqrt{x^2 + y^2}$, as indicated. Then again the norm of A by the Pythagoras theorem would be

$$\sqrt{w^2 + z^2} = \sqrt{x^2 + y^2 + z^2}.$$

Thus when $n = 3$, our definition of norm is compatible with the geometry of the Pythagoras theorem.

In terms of coordinates, $A = (a_1, \ldots, a_n)$ we see that

$$\|A\| = \sqrt{a_1^2 + \cdots + a_n^2}.$$

If $A \neq O$, then $\|A\| \neq 0$ because some coordinate $a_i \neq 0$, so that $a_i^2 > 0$, and hence $a_1^2 + \cdots + a_n^2 > 0$, so $\|A\| \neq 0$.

Observe that for any vector A we have

$$\boxed{\|A\| = \|-A\|.}$$

This is due to the fact that

$$(-a_1)^2 + \cdots + (-a_n)^2 = a_1^2 + \cdots + a_n^2,$$

because $(-1)^2 = 1$. Of course, this is as it should be from the picture:

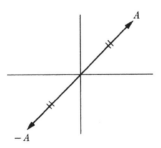

Figure 17

Recall that A and $-A$ are said to have **opposite direction**. However, they have the same norm (magnitude, as is sometimes said when speaking of vectors).

Let A, B be two points. We define the **distance** between A and B to be

$$\|A - B\| = \sqrt{(A - B) \cdot (A - B)}.$$

This definition coincides with our geometric intuition when A, B are points in the plane (Fig. 18). It is the same thing as the length of the located vector $\overrightarrow{AB}$ or the located vector $\overrightarrow{BA}$.

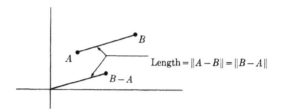

Figure 18

Example 3. Let $A = (-1, 2)$ and $B = (3, 4)$. Then the length of the located vector $\overrightarrow{AB}$ is $\|B - A\|$. But $B - A = (4, 2)$. Thus

$$\|B - A\| = \sqrt{16 + 4} = \sqrt{20}.$$

In the picture, we see that the horizontal side has length 4 and the vertical side has length 2. Thus our definitions reflect our geometric intuition derived from Pythagoras.

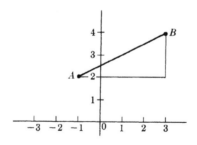

Figure 19

Let P be a point in the plane, and let a be a number > 0. The set of points X such that

$$\|X - P\| < a$$

will be called the **open disc** of radius a centered at P. The set of points X such that

$$\|X - P\| \leqq a$$

will be called the **closed disc** of radius a and center P. The set of points X such that

$$\|X - P\| = a$$

is called the circle of radius a and center P. These are illustrated in Fig. 20.

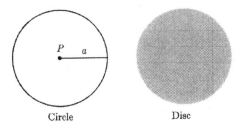

Circle   Disc

Figure 20

In 3-dimensional space, the set of points X such that

$$\|X - P\| < a$$

will be called the **open ball** of radius a and center P. The set of points X such that

$$\|X - P\| \leq a$$

will be called the **closed ball** of radius a and center P. The set of points X such that

$$\|X - P\| = a$$

will be called the **sphere** of radius a and center P. In higher dimensional space, one uses this same terminology of ball and sphere.

Figure 21 illustrates a sphere and a ball in 3-space.

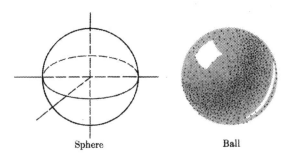

Sphere   Ball

Figure 21

The sphere is the outer shell, and the ball consists of the region inside the shell. The open ball consists of the region inside the shell excluding the shell itself. The closed ball consists of the region inside the shell *and* the shell itself.

From the geometry of the situation, it is also reasonable to expect that if $c > 0$, then $\|cA\| = c\|A\|$, i.e. if we stretch a vector A by multiplying by a positive number c, then the length stretches also by that amount. We verify this formally using our definition of the length.

Theorem 4.1 *Let x be a number. Then*

$$\|xA\| = |x|\,\|A\|$$

(absolute value of x times the norm of A).

Proof. By definition, we have

$$\|xA\|^2 = (xA)\cdot(xA),$$

which is equal to

$$x^2(A\cdot A)$$

by the properties of the scalar product. Taking the square root now yields what we want.

Let S_1 be the sphere of radius 1, centered at the origin. Let a be a number > 0. If X is a point of the sphere S_1, then aX is a point of the sphere of radius a, because

$$\|aX\| = a\|X\| = a.$$

In this manner, we get all points of the sphere of radius a. (Proof?) Thus the sphere of radius a is obtained by stretching the sphere of radius 1, through multiplication by a.

A similar remark applies to the open and closed balls of radius a, they being obtained from the open and closed balls of radius 1 through multiplication by a.

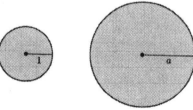

Disc o

Figure 22

We shall say that a vector E is a **unit** vector if $\|E\| = 1$. Given any vector A, let $a = \|A\|$. If $a \neq 0$, then

$$\frac{1}{a} A$$

is a unit vector, because

$$\left\| \frac{1}{a} A \right\| = \frac{1}{a} a = 1.$$

We say that two vectors A, B (neither of which is O) have the **same direction** if there is a number $c > 0$ such that $cA = B$. In view of this definition, we see that the vector

$$\frac{1}{\|A\|} A$$

is a unit vector in the direction of A (provided $A \neq O$).

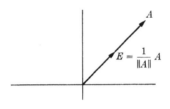

Figure 23

If E is the unit vector in the direction of A, and $\|A\| = a$, then

$$A = aE.$$

Example 4. Let $A = (1, 2, -3)$. Then $\|A\| = \sqrt{14}$. Hence the unit vector in the direction of A is the vector

$$E = \left(\frac{1}{\sqrt{14}}, \frac{2}{\sqrt{14}}, \frac{-3}{\sqrt{14}} \right).$$

Warning. There are as many unit vectors as there are directions. The three **standard unit vectors** in 3-space, namely

$$E_1 = (1, 0, 0), \qquad E_2 = (0, 1, 0), \qquad E_3 = (0, 0, 1)$$

are merely the three unit vectors in the directions of the coordinate axes.

We are also in the position to justify our definition of perpendicularity. Given A, B in the plane, the condition that

$$\|A + B\| = \|A - B\|$$

(illustrated in Fig. 24(b)) coincides with the geometric property that A should be perpendicular to B.

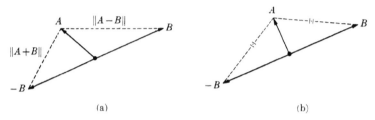

(a)   (b)

Figure 24

We shall prove:

$$\|A + B\| = \|A - B\| \text{ if and only if } A \cdot B = 0.$$

Let $\Leftrightarrow$ denote "if and only if". Then

$$\|A + B\| = \|A - B\| \Leftrightarrow \|A + B\|^2 = \|A - B\|^2$$
$$\Leftrightarrow A^2 + 2A \cdot B + B^2 = A^2 - 2A \cdot B + B^2$$
$$\Leftrightarrow 4A \cdot B = 0$$
$$\Leftrightarrow A \cdot B = 0.$$

This proves what we wanted.

General Pythagoras theorem. *If A and B are perpendicular, then*

$$\|A + B\|^2 = \|A\|^2 + \|B\|^2.$$

The theorem is illustrated on Fig. 25.

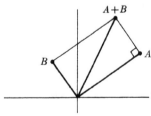

Figure 25

To prove this, we use the definitions, namely

$$\|A + B\|^2 = (A + B) \cdot (A + B) = A^2 + 2A \cdot B + B^2$$
$$= \|A\|^2 + \|B\|^2,$$

because $A \cdot B = 0$, and $A \cdot A = \|A\|^2$, $B \cdot B = \|B\|^2$ by definition.

Remark. If A is perpendicular to B, and x is any number, then A is also perpendicular to xB because

$$A \cdot xB = xA \cdot B = 0.$$

We shall now use the notion of perpendicularity to derive the notion of **projection**. Let A, B be two vectors and $B \neq O$. Let P be the point on the line through $\overrightarrow{OB}$ such that $\overrightarrow{PA}$ is perpendicular to $\overrightarrow{OB}$, as shown on Fig. 26(a).

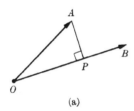
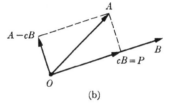

(a)   (b)

Figure 26

We can write

$$P = cB$$

for some number c. We want to find this number c explicitly in terms of A and B. The condition $\overrightarrow{PA} \perp \overrightarrow{OB}$ means that

$$A - P \text{ is perpendicular to } B,$$

and since $P = cB$ this means that

$$(A - cB) \cdot B = 0,$$

in other words,

$$A \cdot B - cB \cdot B = 0.$$

We can solve for c, and we find $A \cdot B = cB \cdot B$, so that

$$c = \frac{A \cdot B}{B \cdot B}.$$

Conversely, if we take this value for c, and then use distributivity, dotting $A - cB$ with B yields 0, so that $A - cB$ is perpendicular to B. Hence we have seen that there is a unique number c such that $A - cB$ is perpendicular to B, and c is given by the above formula.

Definition. The **component** of A along B is the number $c = \dfrac{A \cdot B}{B \cdot B}$.

The **projection** of A along B is the vector $cB = \dfrac{A \cdot B}{B \cdot B} B$.

Example 5. Suppose

$$B = E_i = (0, \ldots, 0, 1, 0, \ldots, 0)$$

is the i-th unit vector, with 1 in the i-th component and 0 in all other components.

$$\textit{If } A = (a_1, \ldots, a_n), \textit{ then } A \cdot E_i = a_i.$$

Thus $A \cdot E_i$ is the ordinary i-th component of A.

More generally, if B is a unit vector, not necessarily one of the E_i, then we have simply

$$c = A \cdot B$$

because $B \cdot B = 1$ by definition of a unit vector.

Example 6. Let $A = (1, 2, -3)$ and $B = (1, 1, 2)$. Then the component of A along B is the **number**

$$c = \frac{A \cdot B}{B \cdot B} = \frac{-3}{6} = -\frac{1}{2}.$$

Hence the projection of A along B is the **vector**

$$cB = (-\tfrac{1}{2}, -\tfrac{1}{2}, -1).$$

Our construction gives an immediate geometric interpretation for the scalar product. Namely, assume $A \neq O$ and look at the angle θ between A and B (Fig. 27). Then from plane geometry we see that

$$\cos \theta = \frac{c\|B\|}{\|A\|},$$

or substituting the value for c obtained above,

$$\boxed{A \cdot B = \|A\| \, \|B\| \cos \theta} \quad \text{and} \quad \boxed{\cos \theta = \frac{A \cdot B}{\|A\| \, \|B\|}.}$$

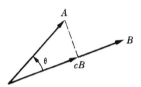

Figure 27

In some treatments of vectors, one takes the relation

$$A \cdot B = \|A\| \, \|B\| \, \cos \theta$$

as definition of the scalar product. This is subject to the following disadvantages, not to say objections:

(a) The four properties of the scalar product **SP 1** through **SP 4** are then by no means obvious.
(b) Even in 3-space, one has to rely on geometric intuition to obtain the cosine of the angle between A and B, and this intuition is less clear than in the plane. In higher dimensional space, it fails even more.
(c) It is extremely hard to work with such a definition to obtain further properties of the scalar product.

Thus we prefer to lay obvious algebraic foundations, and then recover very simply all the properties. We used plane geometry to see the expression

$$A \cdot B = \|A\| \, \|B\| \, \cos \theta.$$

After working out some examples, we shall prove the inequality which allows us to justify this in n-space.

Example 7. Let $A = (1, 2, -3)$ and $B = (2, 1, 5)$. Find the cosine of the angle θ between A and B.
By definition,

$$\cos \theta = \frac{A \cdot B}{\|A\| \, \|B\|} = \frac{2 + 2 - 15}{\sqrt{14} \, \sqrt{30}} = \frac{-11}{\sqrt{420}}.$$

Example 8. Find the cosine of the angle between the two located vectors $\overrightarrow{PQ}$ and $\overrightarrow{PR}$ where

$$P = (1, 2, -3), \qquad Q = (-2, 1, 5), \qquad R = (1, 1, -4).$$

The picture looks like this:

Figure 28

We let

$$A = Q - P = (-3, -1, 8) \quad \text{and} \quad B = R - P = (0, -1, -1).$$

Then the angle between $\overrightarrow{PQ}$ and $\overrightarrow{PR}$ is the same as that between A and B. Hence its cosine is equal to

$$\cos \theta = \frac{A \cdot B}{\|A\| \, \|B\|} = \frac{0 + 1 - 8}{\sqrt{74} \sqrt{2}} = \frac{-7}{\sqrt{74} \sqrt{2}}.$$

We shall prove further properties of the norm and scalar product using our results on perpendicularity. First note a special case. If

$$E_i = (0, \ldots, 0, 1, 0, \ldots, 0)$$

is the i-th unit vector of $\mathbf{R}^n$, and

$$A = (a_1, \ldots, a_n),$$

then

$$A \cdot E_i = a_i$$

is the i-th component of A, i.e. the component of A along E_i. We have

$$|a_i| = \sqrt{a_i^2} \leq \sqrt{a_1^2 + \cdots + a_n^2} = \|A\|,$$

so that the absolute value of each component of A is at most equal to the length of A.

We don't have to deal only with the special unit vector as above. Let E be any unit vector, that is a vector of norm 1. Let c be the component of A along E. We saw that

$$c = A \cdot E.$$

Then $A - cE$ is perpendicular to E, and

$$A = A - cE + cE.$$

Then $A - cE$ is also perpendicular to cE, and by the Pythagoras theorem, we find

$$\|A\|^2 = \|A - cE\|^2 + \|cE\|^2 = \|A - cE\|^2 + c^2.$$

Thus we have the inequality $c^2 \leq \|A\|^2$, and $|c| \leq \|A\|$.

In the next theorem, we generalize this inequality to a dot product $A \cdot B$ when B is not necessarily a unit vector.

Theorem 4.2. *Let A, B be two vectors in $\mathbf{R}^n$. Then*

$$|A \cdot B| \leq \|A\| \, \|B\|.$$

Proof. If $B = O$, then both sides of the inequality are equal to 0, and so our assertion is obvious. Suppose that $B \neq O$. Let c be the component of A along B, so $c = (A \cdot B)/(B \cdot B)$. We write

$$A = A - cB + cB.$$

By Pythagoras,

$$\|A\|^2 = \|A - cB\|^2 + \|cB\|^2 = \|A - cB\|^2 + c^2\|B\|^2.$$

Hence $c^2\|B\|^2 \leq \|A\|^2$. But

$$c^2\|B\|^2 = \frac{(A \cdot B)^2}{(B \cdot B)^2}\|B\|^2 = \frac{|A \cdot B|^2}{\|B\|^4}\|B\|^2 = \frac{|A \cdot B|^2}{\|B\|^2}.$$

Therefore

$$\frac{|A \cdot B|^2}{\|B\|^2} \leq \|A\|^2.$$

Multiply by $\|B\|^2$ and take the square root to conclude the proof.

In view of Theorem 4.2, we see that for vectors A, B in n-space, the number

$$\frac{A \cdot B}{\|A\| \, \|B\|}$$

has absolute value ≤ 1. Consequently,

$$-1 \leq \frac{A \cdot B}{\|A\| \, \|B\|} \leq 1,$$

and there exists a unique angle θ such that $0 \le \theta \le \pi$, and such that

$$\cos\theta = \frac{A \cdot B}{\|A\| \, \|B\|}.$$

We define this angle to be the **angle between A and B**.

The inequality of Theorem 4.2 is known as the **Schwarz inequality**.

Theorem 4.3. *Let A, B be vectors. Then*

$$\|A + B\| \le \|A\| + \|B\|.$$

Proof. Both sides of this inequality are positive or 0. Hence it will suffice to prove that their squares satisfy the desired inequality, in other words,

$$(A + B) \cdot (A + B) \le (\|A\| + \|B\|)^2.$$

To do this, we consider

$$(A + B) \cdot (A + B) = A \cdot A + 2A \cdot B + B \cdot B.$$

In view of our previous result, this satisfies the inequality

$$\le \|A\|^2 + 2\|A\| \, \|B\| + \|B\|^2,$$

and the right-hand side is none other than

$$(\|A\| + \|B\|)^2.$$

Our theorem is proved.

Theorem 4.3 is known as the **triangle inequality**. The reason for this is that if we draw a triangle as in Fig. 29, then Theorem 4.3 expresses the fact that the length of one side is $\le$ the sum of the lengths of the other two sides.

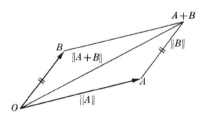

Figure 29

Remark. All the proofs do not use coordinates, only properties **SP 1** through **SP 4** of the dot product. Hence they remain valid in more general situations, see Chapter VI. In n-space, they give us inequalities which are by no means obvious when expressed in terms of coordinates. For instance, the Schwarz inequality reads, in terms of coordinates:

$$|a_1 b_1 + \cdots + a_n b_n| \leqq (a_1^2 + \cdots + a_n^2)^{1/2}(b_1^2 + \cdots + b_n^2)^{1/2}.$$

Just try to prove this directly, without the "geometric" intuition of Pythagoras, and see how far you get.

Exercises I, §4

1. Find the norm of the vector A in the following cases.
   (a) $A = (2, -1)$, $B = (-1, 1)$
   (b) $A = (-1, 3)$, $B = (0, 4)$
   (c) $A = (2, -1, 5)$, $B = (-1, 1, 1)$
   (d) $A = (-1, -2, 3)$, $B = (-1, 3, -4)$
   (e) $A = (\pi, 3, -1)$, $B = (2\pi, -3, 7)$
   (f) $A = (15, -2, 4)$, $B = (\pi, 3, -1)$

2. Find the norm of vector B in the above cases.

3. Find the projection of A along B in the above cases.

4. Find the projection of B along A in the above cases.

5. Find the cosine between the following vectors A and B.
   (a) $A = (1, -2)$ and $B = (5, 3)$
   (b) $A = (-3, 4)$ and $B = (2, -1)$
   (c) $A = (1, -2, 3)$ and $B = (-3, 1, 5)$
   (d) $A = (-2, 1, 4)$ and $B = (-1, -1, 3)$
   (e) $A = (-1, 1, 0)$ and $B = (2, 1, -1)$

6. Determine the cosine of the angles of the triangle whose vertices are
   (a) $(2, -1, 1)$, $(1, -3, -5)$, $(3, -4, -4)$.
   (b) $(3, 1, 1)$, $(-1, 2, 1)$, $(2, -2, 5)$.

7. Let $A_1, \ldots, A_r$ be non-zero vectors which are mutually perpendicular, in other words $A_i \cdot A_j = 0$ if $i \neq j$. Let $c_1, \ldots, c_r$ be numbers such that

$$c_1 A_1 + \cdots + c_r A_r = O.$$

   Show that all $c_i = 0$.

8. For any vectors A, B, prove the following relations:
   (a) $\|A + B\|^2 + \|A - B\|^2 = 2\|A\|^2 + 2\|B\|^2$.
   (b) $\|A + B\|^2 = \|A\|^2 + \|B\|^2 + 2A \cdot B$.
   (c) $\|A + B\|^2 - \|A - B\|^2 = 4A \cdot B$.
   Interpret (a) as a "parallelogram law".

9. Show that if θ is the angle between A and B, then

$$\|A - B\|^2 = \|A\|^2 + \|B\|^2 - 2\|A\| \, \|B\| \, \cos \theta.$$

10. Let A, B, C be three non-zero vectors. If $A \cdot B = A \cdot C$, show by an example that we do not necessarily have $B = C$.

I, §5. Parametric Lines

We define the **parametric equation** or **parametric representation** of a straight line passing through a point P in the direction of a vector $A \neq O$ to be

$$X = P + tA,$$

where t runs through all numbers (Fig. 30).

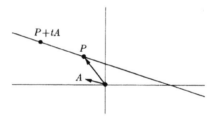

Figure 30

When we give such a parametric representation, we may think of a bug starting from a point P at time $t = 0$, and moving in the direction of A. At time t, the bug is at the position $P + tA$. Thus we may interpret physically the parametric representation as a description of motion, in which A is interpreted as the velocity of the bug. At a given time t, the bug is at the point.

$$X(t) = P + tA,$$

which is called the **position** of the bug at time t.

This parametric representation is also useful to describe the set of points lying on the line segment between two given points. Let P, Q be two points. Then the **segment** between P and Q consists of all the points

$$S(t) = P + t(Q - P) \qquad \text{with} \qquad 0 \leq t \leq 1.$$

Indeed, $\overrightarrow{O(Q - P)}$ is a vector having the same direction as $\overrightarrow{PQ}$, as shown on Fig. 31.

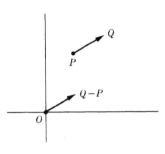

Figure 31

When $t = 0$, we have $S(0) = P$, so at time $t = 0$ the bug is at P. When $t = 1$, we have

$$S(1) = P + (Q - P) = Q,$$

so when $t = 1$ the bug is at Q. As t goes from 0 to 1, the bug goes from P to Q.

Example 1. Let $P = (1, -3, 4)$ and $Q = (5, 1, -2)$. Find the coordinates of the point which lies one third of the distance from P to Q.

Let $S(t)$ as above be the parametric representation of the segment from P to Q. The desired point is $S(1/3)$, that is:

$$S\left(\frac{1}{3}\right) = P + \frac{1}{3}(Q - P) = (1, -3, 4) + \frac{1}{3}(4, 4, -6)$$

$$= \left(\frac{7}{3}, \frac{-5}{3}, 2\right).$$

Warning. The desired point in the above example is *not* given by

$$\frac{P + Q}{3}.$$

Example 2. Find a parametric representation for the line passing through the two points $P = (1, -3, 1)$ and $Q = (-2, 4, 5)$.

We first have to find a vector in the direction of the line. We let

$$A = P - Q,$$

so

$$A = (3, -7, -4).$$

The parametric representation of the line is therefore

$$X(t) = P + tA = (1, -3, 1) + t(3, -7, -4).$$

Remark. It would be equally correct to give a parametric representation of the line as

$$Y(t) = P + tB \qquad \text{where} \qquad B = Q - P.$$

Interpreted in terms of the moving bug, however, one parametrization gives the position of a bug moving in one direction along the line, starting from P at time $t = 0$, while the other parametrization gives the position of another bug moving in the **opposite** direction along the line, also starting from P at time $t = 0$.

We shall now discuss the relation between a parametric representation and the ordinary equation of a line in the plane.

Suppose that we work in the plane, and write the coordinates of a point X as (x, y). Let $P = (p, q)$ and $A = (a, b)$. Then in terms of the coordinates, we can write

$$x = p + ta, \qquad y = q + tb.$$

We can then eliminate t and obtain the usual equation relating x and y.

Example 3. Let $P = (2, 1)$ and $A = (-1, 5)$. Then the parametric representation of the line through P in the direction of A gives us

$$(*) \qquad\qquad x = 2 - t, \qquad y = 1 + 5t.$$

Multiplying the first equation by 5 and adding yields

$$(**) \qquad\qquad 5x + y = 11,$$

which is the familiar equation of a line.

This elimination of t shows that every pair (x, y) which satisfies the parametric representation $(*)$ for some value of t also satisfies equation $(**)$. Conversely, suppose we have a pair of numbers (x, y) satisfying $(**)$. Let $t = 2 - x$. Then

$$y = 11 - 5x = 11 - 5(2 - t) = 1 + 5t.$$

Hence there exists some value of t which satisfies equation (∗). Thus we have proved that the pairs (x, y) which are solutions of (∗∗) are exactly the same pairs of numbers as those obtained by giving arbitrary values for t in (∗). Thus the straight line can be described parametrically as in (∗) or in terms of its usual equation (∗∗). Starting with the ordinary equation

$$5x + y = 11,$$

we let $t = 2 - x$ in order to recover the specific parametrization of (∗).

When we parametrize a straight line in the form

$$X = P + tA,$$

we have of course infinitely many choices for P on the line, and also infinitely many choices for A, differing by a scalar multiple. We can always select at least one. Namely, given an equation

$$ax + by = c$$

with numbers a, b, c, suppose that $a \neq 0$. We use y as parameter, and let

$$y = t.$$

Then we can solve for x, namely

$$x = \frac{c}{a} - \frac{b}{a} t.$$

Let $P = (c/a, 0)$ and $A = (-b/a, 1)$. We see that an arbitrary point (x, y) satisfying the equation

$$ax + by = c$$

can be expressed parametrically, namely

$$(x, y) = P + tA.$$

In higher dimensions, starting with a parametric representation

$$X = P + tA,$$

we cannot eliminate t, and thus the parametric representation is the only one available to describe a straight line.

Exercises I, §5

1. Find a parametric representation for the line passing through the following pairs of points.
   (a) $P_1 = (1, 3, -1)$ and $P_2 = (-4, 1, 2)$
   (b) $P_1 = (-1, 5, 3)$ and $P_2 = (-2, 4, 7)$

Find a parametric representation for the line passing through the following points.

2. $(1, 1, -1)$ and $(-2, 1, 3)$ 3. $(-1, 5, 2)$ and $(3, -4, 1)$

4. Let $P = (1, 3, -1)$ and $Q = (-4, 5, 2)$. Determine the coordinates of the following points:
   (a) The midpoint of the line segment between P and Q.
   (b) The two points on this line segment lying one-third and two-thirds of the way from P to Q.
   (c) The point lying one-fifth of the way from P to Q.
   (d) The point lying two-fifths of the way from P to Q.

5. If P, Q are two arbitrary points in n-space, give the general formula for the midpoint of the line segment between P and Q.

I, §6. Planes

We can describe planes in 3-space by an equation analogous to the single equation of the line. We proceed as follows.

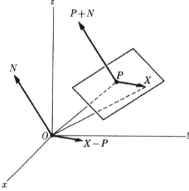

Figure 32

Let P be a point in 3-space and consider a located vector $\overrightarrow{ON}$. We define the **plane passing through P perpendicular to $\overrightarrow{ON}$** to be the collection of all points X such that the located vector $\overrightarrow{PX}$ is perpendicular to $\overrightarrow{ON}$. According to our definitions, this amounts to the condition

$$(X - P) \cdot N = 0,$$

which can also be written as

$$X \cdot N = P \cdot N.$$

We shall also say that this plane is the one perpendicular to N, and consists of all vectors X such that $X - P$ is perpendicular to N. We have drawn a typical situation in 3-spaces in Fig. 32.

Instead of saying that N is **perpendicular** to the plane, one also says that N is **normal** to the plane.

Let t be a number $\neq 0$. Then the set of points X such that

$$(X - P) \cdot N = 0$$

coincides with the set of points X such that

$$(X - P) \cdot tN = 0.$$

Thus we may say that our plane is the plane passing through P and perpendicular to the **line** in the direction of N. To find the equation of the plane, we could use any vector tN (with $t \neq 0$) instead of N.

Example 1. Let

$$P = (2, 1, -1) \qquad \text{and} \qquad N = (-1, 1, 3).$$

Let $X = (x, y, z)$. Then

$$X \cdot N = (-1)x + y + 3z.$$

Therefore the equation of the plane passing through P and perpendicular to N is

$$-x + y + 3z = -2 + 1 - 3$$

or

$$-x + y + 3z = -4.$$

Observe that in 2-space, with $X = (x, y)$, the formulas lead to the equation of the line in the ordinary sense.

Example 2. The equation of the line in the (x, y)-plane, passing through $(4, -3)$ and perpendicular to $(-5, 2)$ is

$$-5x + 2y = -20 - 6 = -26.$$

We are now in position to interpret the coefficients $(-5, 2)$ of x and y in this equation. They give rise to a vector perpendicular to the line. **In any equation**

$$ax + by = c$$

the vector (a, b) is perpendicular to the line determined by the equation. Similarly, in 3-space, the vector (a, b, c) is perpendicular to the plane determined by the equation

$$ax + by + cz = d.$$

Example 3. The plane determined by the equation

$$2x - y + 3z = 5$$

is perpendicular to the vector $(2, -1, 3)$. If we want to find a point in that plane, we of course have many choices. We can give arbitrary values to x and y, and then solve for z. To get a concrete point, let $x = 1$, $y = 1$. Then we solve for z, namely

$$3z = 5 - 2 + 1 = 4,$$

so that $z = \frac{4}{3}$. Thus

$$(1, 1, \tfrac{4}{3})$$

is a point in the plane.

In n-space, the equation $X \cdot N = P \cdot N$ is said to be the equation of a **hyperplane.** For example,

$$3x - y + z + 2w = 5$$

is the equation of a hyperplane in 4-space, perpendicular to $(3, -1, 1, 2)$.

Two vectors A, B are said to be parallel if there exists a number $c \neq 0$ such that $cA = B$. Two lines are said to be **parallel** if, given two distinct points P_1, Q_1 on the first line and P_2, Q_2 on the second, the vectors

$$P_1 - Q_1$$

and

$$P_2 - Q_2$$

are parallel.

Two planes are said to be **parallel** (in 3-space) if their normal vectors are parallel. They are said to be **perpendicular** if their normal vectors are perpendicular. The **angle** between two planes is defined to be the angle between their normal vectors.

Example 4. Find the cosine of the angle θ between the planes.

$$2x - y + z = 0,$$
$$x + 2y - z = 1.$$

This cosine is the cosine of the angle between the vectors

$$A = (2, -1, 1) \qquad \text{and} \qquad B = (1, 2, -1).$$

Therefore

$$\cos \theta = \frac{A \cdot B}{\|A\| \, \|B\|} = -\frac{1}{6}.$$

Example 5. Let

$$Q = (1, 1, 1) \qquad \text{and} \qquad P = (1, -1, 2).$$

Let

$$N = (1, 2, 3)$$

Find the point of intersection of the line through P in the direction of N, and the plane through Q perpendicular to N.

The parametric representation of the line through P in the direction of N is

(1) $X = P + tN.$

The equation of the plane through Q perpendicular to N is

(2) $(X - Q) \cdot N = 0.$

We visualize the line and plane as follows:

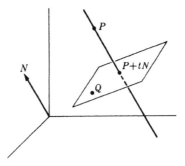

Figure 33

We must find the value of t such that the vector X in (1) also satisfies (2), that is

$$(P + tN - Q) \cdot N = 0,$$

or after using the rules of the dot product,

$$(P - Q) \cdot N + tN \cdot N = 0.$$

Solving for t yields

$$t = \frac{(Q - P) \cdot N}{N \cdot N} = \frac{1}{14}.$$

Thus the desired point of intersection is

$$P + tN = (1, -1, 2) + \tfrac{1}{14}(1, 2, 3) = (\tfrac{15}{14}, -\tfrac{12}{14}, \tfrac{31}{14}).$$

Example 6. Find the equation of the plane passing through the three points

$$P_1 = (1, 2, -1). \qquad P_2 = (-1, 1, 4), \qquad P_3 = (1, 3, -2).$$

We visualize schematically the three points as follows:

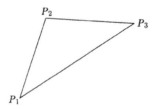

Figure 34

Then we find a vector N perpendicular to $\overrightarrow{P_1P_2}$ and $\overrightarrow{P_1P_3}$, or in other words, perpendicular to $P_2 - P_1$ and $P_3 - P_1$. We have

$$P_2 - P_1 = (-2, -1, +5),$$
$$P_3 - P_1 = (0, 1, -1).$$

Let $N = (a, b, c)$. We must solve

$$N \cdot (P_2 - P_1) = 0 \qquad \text{and} \qquad N \cdot (P_3 - P_1) = 0,$$

in other words,

$$-2a - b + 5c = 0,$$
$$b - c = 0.$$

We take $b = c = 1$ and solve for $a = 2$. Then

$$N = (2, 1, 1)$$

satisfies our requirements. The plane perpendicular to N, passing through P_1 is the desired plane. Its equation is therefore $X \cdot N = P_1 \cdot N$, that is

$$2x + y + z = 2 + 2 - 1 = 3.$$

Distance between a point and a plane. Consider a plane defined by the equation

$$(X - P) \cdot N = 0,$$

and let Q be an arbitrary point. We wish to find a formula for the distance between Q and the plane. By this we mean the length of the segment from Q to the point of intersection of the perpendicular line to the plane through Q, as on the figure. We let Q' be this point of intersection.

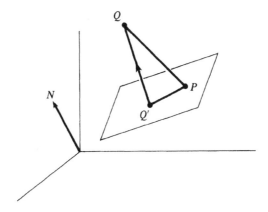

Figure 35

From the geometry, we have:

length of the segment $\overline{QQ'}$ = length of the projection of $\overline{QP}$ on $\overline{QQ'}$.

We can express the length of this projection in terms of the dot product as follows. A unit vector in the direction of N, which is perpendicular to the plane, is given by $N/\|N\|$. Then

length of the projection of $\overline{QP}$ on $\overline{QQ'}$

$$= \text{norm of the projection of } Q - P \text{ on } N/\|N\|$$

$$= \left| (Q - P) \cdot \frac{N}{\|N\|} \right|.$$

This can also be written in the form:

$$\text{distance between } Q \text{ and the plane} = \frac{|(Q - P) \cdot N|}{\|N\|}.$$

Example 7. Let

$$Q = (1, 3, 5), \qquad P = (-1, 1, 7) \qquad \text{and} \qquad N = (-1, 1, -1).$$

The equation of the plane is

$$-x + y - z = -5.$$

We find $\|N\| = \sqrt{3}$,

$$Q - P = (2, 2, -2) \qquad \text{and} \qquad (Q - P) \cdot N = -2 + 2 + 2 = 2.$$

Hence the distance between Q and the plane is $2/\sqrt{3}$.

Exercises I, §6

1. Show that the lines $2x + 3y = 1$ and $5x - 5y = 7$ are not perpendicular.

2. Let $y = mx + b$ and $y = m'x + c$ be the equations of two lines in the plane. Write down vectors perpendicular to these lines. Show that these vectors are perpendicular to each other if and only if $mm' = -1$.

Find the equation of the line in 2-space, perpendicular to N and passing through P, for the following values of N and P.

3. $N = (1, -1)$, $P = (-5, 3)$ 4. $N = (-5, 4)$, $P = (3, 2)$

5. Show that the lines

$$3x - 5y = 1, \qquad 2x + 3y = 5$$

are not perpendicular.

6. Which of the following pairs of lines are perpendicular?
   (a) $3x - 5y = 1$ and $2x + y = 2$
   (b) $2x + 7y = 1$ and $x - y = 5$
   (c) $3x - 5y = 1$ and $5x + 3y = 7$
   (d) $-x + y = 2$ and $x + y = 9$

7. Find the equation of the plane perpendicular to the given vector N and passing through the given point P.
   (a) $N = (1, -1, 3)$, $P = (4, 2, -1)$
   (b) $N = (-3, -2, 4)$, $P = (2, \pi, -5)$
   (c) $N = (-1, 0, 5)$, $P = (2, 3, 7)$

8. Find the equation of the plane passing through the following three points.
   (a) $(2, 1, 1)$, $(3, -1, 1)$, $(4, 1, -1)$
   (b) $(-2, 3, -1)$, $(2, 2, 3)$, $(-4, -1, 1)$
   (c) $(-5, -1, 2)$, $(1, 2, -1)$, $(3, -1, 2)$

9. Find a vector perpendicular to $(1, 2, -3)$ and $(2, -1, 3)$, and another vector perpendicular to $(-1, 3, 2)$ and $(2, 1, 1)$.

10. Find a vector parallel to the line of intersection of the two planes

$$2x - y + z = 1, \qquad 3x + y + z = 2.$$

11. Same question for the planes,

$$2x + y + 5z = 2, \qquad 3x - 2y + z = 3.$$

12. Find a parametric representation for the line of intersection of the planes of Exercises 10 and 11.

13. Find the cosine of the angle between the following planes:
   (a) $x + y + z = 1$ (b) $2x + 3y - z = 2$
       $x - y - z = 5$ $x - y + z = 1$
   (c) $x + 2y - z = 1$ (d) $2x + y + z = 3$
       $-x + 3y + z = 2$ $-x - y + z = \pi$

14. (a) Let $P = (1, 3, 5)$ and $A = (-2, 1, 1)$. Find the intersection of the line through P in the direction of A, and the plane $2x + 3y - z = 1$.
    (b) Let $P = (1, 2, -1)$. Find the point of intersection of the plane

$$3x - 4y + z = 2,$$

with the line through P, perpendicular to that plane.

15. Let $Q = (1, -1, 2)$, $P = (1, 3, -2)$, and $N = (1, 2, 2)$. Find the point of the intersection of the line through P in the direction of N, and the plane through Q perpendicular to N.

16. Find the distance between the indicated point and plane.
   (a) $(1, 1, 2)$ and $3x + y - 5z = 2$
   (b) $(-1, 3, 2)$ and $2x - 4y + z = 1$
   (c) $(3, -2, 1)$ and the yz-plane
   (d) $(-3, -2, 1)$ and the yz-plane

CHAPTER II

Matrices and Linear Equations

You have met linear equations in elementary school. Linear equations are simply equations like

$$2x + y + \ z = 1,$$
$$5x - y + 7z = 0.$$

You have learned to solve such equations by the successive elimination of the variables. In this chapter, we shall review the theory of such equations, dealing with equations in n variables, and interpreting our results from the point of view of vectors. Several geometric interpretations for the solutions of the equations will be given.

The first chapter is used here very little, and can be entirely omitted if you know only the definition of the dot product between two n-tuples. The multiplication of matrices will be formulated in terms of such a product. One geometric interpretation for the solutions of homogeneous equations will however rely on the fact that the dot product between two vectors is 0 if and only if the vectors are perpendicular, so if you are interested in this interpretation, you should refer to the section in Chapter I where this is explained.

II, §1. Matrices

We consider a new kind of object, matrices.

Let n, m be two integers ≥ 1. An array of numbers

$$\begin{pmatrix} a_{11} & a_{12} & a_{13} & \cdots & a_{1n} \\ a_{21} & a_{22} & a_{23} & \cdots & a_{2n} \\ \vdots & \vdots & \vdots & & \vdots \\ a_{m1} & a_{m2} & a_{m3} & \cdots & a_{mn} \end{pmatrix}$$

is called a **matrix**. We can abbreviate the notation for this matrix by writing it (a_{ij}), $i = 1,\ldots,m$ and $j = 1,\ldots,n$. We say that it is an m by n matrix, or an $m \times n$ matrix. The matrix has m **rows** and n **columns**. For instance, the first column is

$$\begin{pmatrix} a_{11} \\ a_{21} \\ \vdots \\ a_{m1} \end{pmatrix}$$

and the second row is $(a_{21}, a_{22},\ldots,a_{2n})$. We call a_{ij} the **ij-entry** or **ij-component** of the matrix.

Look back at Chapter I, §1. The example of 7-space taken from economics gives rise to a 7×7 matrix (a_{ij}) $(i, j = 1,\ldots,7)$, if we define a_{ij} to be the amount spent by the i-th industry on the j-th industry. Thus keeping the notation of that example, if $a_{25} = 50$, this means that the auto industry bought 50 million dollars worth of stuff from the chemical industry during the given year.

Example 1. The following is a 2×3 matrix:

$$\begin{pmatrix} 1 & 1 & -2 \\ -1 & 4 & -5 \end{pmatrix}.$$

It has two rows and three columns.

The rows are $(1, 1, -2)$ and $(-1, 4, -5)$. The columns are

$$\begin{pmatrix} 1 \\ -1 \end{pmatrix}, \quad \begin{pmatrix} 1 \\ 4 \end{pmatrix}, \quad \begin{pmatrix} -2 \\ -5 \end{pmatrix}.$$

Thus the rows of a matrix may be viewed as n-tuples, and the columns may be viewed as vertical m-tuples. A vertical m-tuple is also called a **column vector**.

A vector $(x_1, \ldots, x_n)$ is a $1 \times n$ matrix. A column vector

$$\begin{pmatrix} x_1 \\ \vdots \\ x_n \end{pmatrix}$$

is an $n \times 1$ matrix.

When we write a matrix in the form (a_{ij}), then i denotes the row and j denotes the column. In Example 1, we have for instance

$$a_{11} = 1, \; a_{23} = -5.$$

A single number (a) may be viewed as a 1×1 matrix.

Let (a_{ij}), $i = 1, \ldots, m$ and $j = 1, \ldots, n$ be a matrix. If $m = n$, then we say that it is a **square** matrix. Thus

$$\begin{pmatrix} 1 & 2 \\ -1 & 0 \end{pmatrix} \quad \text{and} \quad \begin{pmatrix} 1 & -1 & 5 \\ 2 & 1 & -1 \\ 3 & 1 & -1 \end{pmatrix}$$

are both square matrices.

We define the **zero matrix** to be the matrix such that $a_{ij} = 0$ for all i, j. It looks like this:

$$\begin{pmatrix} 0 & 0 & 0 & \cdots & 0 \\ 0 & 0 & 0 & \cdots & 0 \\ \vdots & \vdots & \vdots & & \vdots \\ 0 & 0 & 0 & \cdots & 0 \end{pmatrix}$$

We shall write it O. We note that we have met so far with the zero number, zero vector, and zero matrix.

We shall now define addition of matrices and multiplication of matrices by numbers.

We define addition of matrices only when they have the same size. Thus let m, n be fixed integers ≥ 1. Let $A = (a_{ij})$ and $B = (b_{ij})$ be two $m \times n$ matrices. We define $A + B$ to be the matrix whose entry in the i-th row and j-th column is $a_{ij} + b_{ij}$. In other words, we add matrices of the same size componentwise.

Example 2. Let

$$A = \begin{pmatrix} 1 & -1 & 0 \\ 2 & 3 & 4 \end{pmatrix} \quad \text{and} \quad B = \begin{pmatrix} 5 & 1 & -1 \\ 2 & 1 & -1 \end{pmatrix}.$$

Then

$$A + B = \begin{pmatrix} 6 & 0 & -1 \\ 4 & 4 & 3 \end{pmatrix}.$$

If A, B are both $1 \times n$ matrices, i.e. n-tuples, then we note that our addition of matrices coincides with the addition which we defined in Chapter I for n-tuples.

If O is the zero matrix, then for any matrix A (of the same size, of course), we have $O + A = A + O = A$.

This is trivially verified. We shall now define the multiplication of a matrix by a number. Let c be a number, and $A = (a_{ij})$ be a matrix. We define cA to be the matrix whose ij-component is ca_{ij}. We write

$$cA = (ca_{ij}).$$

Thus we multiply each component of A by c.

Example 3. Let A, B be as in Example 2. Let $c = 2$. Then

$$2A = \begin{pmatrix} 2 & -2 & 0 \\ 4 & 6 & 8 \end{pmatrix} \quad \text{and} \quad 2B = \begin{pmatrix} 10 & 2 & -2 \\ 4 & 2 & -2 \end{pmatrix}.$$

We also have

$$(-1)A = -A = \begin{pmatrix} -1 & 1 & 0 \\ -2 & -3 & -4 \end{pmatrix}.$$

In general, for any matrix $A = (a_{ij})$ we let $-A$ (minus A) be the matrix $(-a_{ij})$. Since we have the relation $a_{ij} - a_{ij} = 0$ for numbers, we also get the relation

$$A + (-A) = O$$

for matrices. The matrix $-A$ is also called the **additive inverse** of A.

We define one more notion related to a matrix. Let $A = (a_{ij})$ be an $m \times n$ matrix. The $n \times m$ matrix $B = (b_{ji})$ such that $b_{ji} = a_{ij}$ is called the **transpose** of A, and is also denoted by ${}^tA$. Taking the transpose of a matrix amounts to changing rows into columns and vice versa. If A is the matrix which we wrote down at the beginning of this section, then ${}^tA$ is the matrix

$$\begin{pmatrix} a_{11} & a_{21} & a_{31} & \cdots & a_{m1} \\ a_{12} & a_{22} & a_{32} & \cdots & a_{m2} \\ \vdots & \vdots & \vdots & & \vdots \\ a_{1n} & a_{2n} & a_{3n} & \cdots & a_{mn} \end{pmatrix}.$$

To take a special case:

$$\text{If } A = \begin{pmatrix} 2 & 1 & 0 \\ 1 & 3 & 5 \end{pmatrix}, \quad \text{then} \quad {}^t A = \begin{pmatrix} 2 & 1 \\ 1 & 3 \\ 0 & 5 \end{pmatrix}.$$

If $A = (2, 1, -4)$ is a *row vector*, then

$${}^t A = \begin{pmatrix} 2 \\ 1 \\ -4 \end{pmatrix}$$

is a *column vector*.

A matrix A which is equal to its transpose, that is $A = {}^t A$, is called **symmetric**. Such a matrix is necessarily a square matrix.

Remark on notation. I have written the transpose sign on the left, because in many situations one considers the inverse of a matrix written A^{-1}, and then it is easier to write ${}^t A^{-1}$ rather than $(A^{-1})^t$ or $(A^t)^{-1}$, which are in fact equal. The mathematical community has no consensus as to where the transpose sign should be placed, on the right or left.

Exercises II, §1

1. Let

$$A = \begin{pmatrix} 1 & 2 & 3 \\ -1 & 0 & 2 \end{pmatrix} \quad \text{and} \quad B = \begin{pmatrix} -1 & 5 & -2 \\ 1 & 1 & -1 \end{pmatrix}.$$

   Find $A + B$, $3B$, $-2B$, $A + 2B$, $2A + B$, $A - B$, $A - 2B$, $B - A$.

2. Let

$$A = \begin{pmatrix} 1 & -1 \\ 2 & 1 \end{pmatrix} \quad \text{and} \quad B = \begin{pmatrix} -1 & 1 \\ 0 & -3 \end{pmatrix}.$$

   Find $A + B$, $3B$, $-2B$, $A + 2B$, $A - B$, $B - A$.

3. (a) Write down the row vectors and column vectors of the matrices A, B in Exercise 1.
   (b) Write down the row vectors and column vectors of the matrices A, B in Exercise 2.

4. (a) In Exercise 1, find ${}^t A$ and ${}^t B$.
   (b) In Exercise 2, find ${}^t A$ and ${}^t B$.

5. If A, B are arbitrary $m \times n$ matrices, show that

$${}^t (A + B) = {}^t A + {}^t B.$$

6. If c is a number, show that $^t(cA) = c\,^tA$.

7. If $A = (a_{ij})$ is a square matrix, then the elements a_{ii} are called the **diagonal** elements. How do the diagonal elements of A and ^tA differ?

8. Find $^t(A + B)$ and $^tA + \,^tB$ in Exercise 2.

9. Find $A + \,^tA$ and $B + \,^tB$ in Exercise 2.

10. (a) Show that for any square matrix, the matrix $A + \,^tA$ is symmetric.
    (b) Define a matrix A to be **skew-symmetric** if $^tA = -A$. Show that for any square matrix A, the matrix $A - \,^tA$ is skew-symmetric.
    (c) If a matrix is skew-symmetric, what can you say about its diagonal elements?

11. Let

$$E_1 = (1,\, 0, \ldots, 0), \quad E_2 = (0,\, 1,\, 0, \ldots, 0), \quad \ldots, \quad E_n = (0, \ldots, 0,\, 1)$$

be the standard unit vectors of $\mathbf{R}^n$. Let $x_1, \ldots, x_n$ be numbers. What is $x_1 E_1 + \cdots + x_n E_n$? Show that if

$$x_1 E_1 + \cdots + x_n E_n = O$$

then $x_i = 0$ for all i.

II, §2. Multiplication of Matrices

We shall now define the product of matrices. Let $A = (a_{ij})$, $i = 1, \ldots, m$ and $j = 1, \ldots, n$ be an $m \times n$ matrix. Let $B = (b_{jk})$, $j = 1, \ldots, n$ and let $k = 1, \ldots, s$ be an $n \times s$ matrix:

$$A = \begin{pmatrix} a_{11} & \cdots & a_{1n} \\ \vdots & & \vdots \\ a_{m1} & \cdots & a_{mn} \end{pmatrix}, \quad B = \begin{pmatrix} b_{11} & \cdots & b_{1s} \\ \vdots & & \vdots \\ b_{n1} & \cdots & b_{ns} \end{pmatrix}.$$

We define the **product** AB to be the $m \times s$ matrix whose ik-coordinate is

$$\sum_{j=1}^{n} a_{ij} b_{jk} = a_{i1} b_{1k} + a_{i2} b_{2k} + \cdots + a_{in} b_{nk}.$$

If $A_1, \ldots, A_m$ are the row vectors of the matrix A, and if $B^1, \ldots, B^s$ are the column vectors of the matrix B, then the ik-coordinate of the product AB is equal to $A_i \cdot B^k$. Thus

$$AB = \begin{pmatrix} A_1 \cdot B^1 & \cdots & A_1 \cdot B^s \\ \vdots & & \vdots \\ A_m \cdot B^1 & \cdots & A_m \cdot B^s \end{pmatrix}.$$

Multiplication of matrices is therefore a generalization of the dot product.

Example. Let

$$A = \begin{pmatrix} 2 & 1 & 5 \\ 1 & 3 & 2 \end{pmatrix}, \qquad B = \begin{pmatrix} 3 & 4 \\ -1 & 2 \\ 2 & 1 \end{pmatrix}.$$

Then AB is a 2×2 matrix, and computations show that

$$AB = \begin{pmatrix} 2 & 1 & 5 \\ 1 & 3 & 2 \end{pmatrix} \begin{pmatrix} 3 & 4 \\ -1 & 2 \\ 2 & 1 \end{pmatrix} = \begin{pmatrix} 15 & 15 \\ 4 & 12 \end{pmatrix}.$$

Example. Let

$$C = \begin{pmatrix} 1 & 3 \\ -1 & -1 \end{pmatrix}.$$

Let A, B be as in Example 1. Then

$$BC = \begin{pmatrix} 3 & 4 \\ -1 & 2 \\ 2 & 1 \end{pmatrix} \begin{pmatrix} 1 & 3 \\ -1 & -1 \end{pmatrix} = \begin{pmatrix} -1 & 5 \\ -3 & -5 \\ 1 & 5 \end{pmatrix}$$

and

$$A(BC) = \begin{pmatrix} 2 & 1 & 5 \\ 1 & 3 & 2 \end{pmatrix} \begin{pmatrix} -1 & 5 \\ -3 & -5 \\ 1 & 5 \end{pmatrix} = \begin{pmatrix} 0 & 30 \\ -8 & 0 \end{pmatrix}.$$

Compute $(AB)C$. What do you find?

If $X = (x_1, \ldots, x_m)$ is a row vector, i.e. a $1 \times m$ matrix, then we can form the product XA, which looks like this:

$$(x_1, \ldots, x_m) \begin{pmatrix} a_{11} & \cdots & a_{1n} \\ \vdots & & \vdots \\ a_{m1} & \cdots & a_{mn} \end{pmatrix} = (y_1, \ldots, y_n),$$

where

$$y_k = x_1 a_{1k} + \cdots + x_m a_{mk}.$$

In this case, XA is a $1 \times n$ matrix, i.e. a row vector.

On the other hand, if X is a column vector,

$$X = \begin{pmatrix} x_1 \\ \vdots \\ x_n \end{pmatrix}$$

then $AX = Y$ where Y is also a column vector, whose coordinates are given by

$$y_i = \sum_{j=1}^{n} a_{ij} x_j = a_{i1} x_1 + \cdots + a_{in} x_n.$$

Visually, the multiplication $AX = Y$ looks like

$$\begin{pmatrix} a_{11} & \cdots & a_{1n} \\ \vdots & & \vdots \\ a_{m1} & \cdots & a_{mn} \end{pmatrix} \begin{pmatrix} x_1 \\ \vdots \\ x_n \end{pmatrix} = \begin{pmatrix} y_1 \\ \vdots \\ y_m \end{pmatrix}.$$

Example. Linear equations. Matrices give a convenient way of writing linear equations. You should already have considered systems of linear equations. For instance, one equation like:

$$3x - 2y + 3z = 1,$$

with three unknowns x, y, z. Or a system of two equations in three unknowns

$$3x - 2y + 3z = 1,$$
(*)
$$-x + 7y - 4z = -5.$$

In this example we let the **matrix of coefficients** be

$$A = \begin{pmatrix} 3 & -2 & 3 \\ -1 & 7 & -4 \end{pmatrix}.$$

Let B be the column vector of the numbers appearing on the right-hand side, so

$$B = \begin{pmatrix} 1 \\ -5 \end{pmatrix}.$$

Let the vector of unknowns be the column vector.

$$X = \begin{pmatrix} x \\ y \\ z \end{pmatrix}.$$

Then you can see that the system of two simultaneous equations can be written in the form

$$AX = B.$$

Example. The first equation of (∗) represents equality of the first component of AX and B; whereas the second equation of (∗) represents equality of the second component of AX and B.

In general, let $A = (a_{ij})$ be an $m \times n$ matrix, and let B be a column vector of size m. Let

$$X = \begin{pmatrix} x_1 \\ x_2 \\ \vdots \\ x_n \end{pmatrix}$$

be a column vector of size n. Then the system of linear equations

$$a_{11}x_1 + \cdots + a_{1n}x_n = b_1,$$
$$a_{21}x_1 + \cdots + a_{2n}x_n = b_2,$$
$$\vdots \qquad \qquad \vdots \quad \vdots$$
$$a_{m1}x_1 + \cdots + a_{mn}x_n = b_m,$$

can be written in the more efficient way

$$\boxed{AX = B,}$$

by the definition of multiplication of matrices. We shall see later how to solve such systems. We say that there are m **equations** and n **unknowns**, or n **variables**.

Example. Markov matrices. A matrix can often be used to represent a practical situation. Suppose we deal with three cities, say Los Angeles, Chicago, and Boston, denoted by LA, Ch, and Bo. Suppose that any given year, some people leave each one of these cities to go to one of the others. The percentages of people leaving and going is given as follows, for each year.

$\frac{1}{4}$ LA goes to Bo	and	$\frac{1}{7}$ LA goes to Ch.
$\frac{1}{5}$ Ch goes to LA	and	$\frac{1}{3}$ Ch goes to Bo.
$\frac{1}{6}$ Bo goes to LA	and	$\frac{1}{8}$ Bo goes to Ch.

Let x_n, y_n, z_n be the populations of LA, Ch, and Bo, respectively, in the n-th year. Then we can express the population in the $(n + 1)$-th year as follows.

In the $(n + 1)$-th year, $\frac{1}{4}$ of the LA population leaves for Boston, and $\frac{1}{7}$ leaves for Chicago. The total fraction leaving LA during the year is therefore

$$\tfrac{1}{4} + \tfrac{1}{7} = \tfrac{11}{28}.$$

Hence the total fraction remaining in LA is

$$1 - \tfrac{11}{28} = \tfrac{17}{28}.$$

Hence the population in LA for the $(n + 1)$-th year is

$$x_{n+1} = \tfrac{17}{28}x_n + \tfrac{1}{5}y_n + \tfrac{1}{6}z_n.$$

Similarly the fraction leaving Chicago each year is

$$\tfrac{1}{5} + \tfrac{1}{3} = \tfrac{8}{15},$$

so the fraction remaining is $\frac{7}{15}$. Finally, the fraction leaving Boston each year is

$$\tfrac{1}{6} + \tfrac{1}{8} = \tfrac{7}{24},$$

so the fraction remaining in Boston is $\frac{17}{24}$. Thus

$$y_{n+1} = \tfrac{1}{7}x_n + \tfrac{7}{15}y_n + \tfrac{1}{8}z_n,$$
$$z_{n+1} = \tfrac{1}{4}x_n + \tfrac{1}{3}y_n + \tfrac{17}{24}z_n.$$

Let A be the matrix

$$A = \begin{pmatrix} \frac{17}{28} & \frac{1}{5} & \frac{1}{6} \\ \frac{1}{7} & \frac{7}{15} & \frac{1}{8} \\ \frac{1}{4} & \frac{1}{3} & \frac{17}{24} \end{pmatrix}.$$

Then we can write down more simply the population shift by the expression

$$X_{n+1} = AX_n \qquad \text{where} \qquad X_n = \begin{pmatrix} x_n \\ y_n \\ z_n \end{pmatrix}.$$

The change from X_n to X_{n+1} is called a **Markov process**. This is due to the special property of the matrix A, all of whose components are ≥ 0, and such that the sum of all the elements in each column is equal to 1. Such a matrix is called a **Markov matrix**.

If A is a square matrix, then we can form the product AA, which will be a square matrix of the same size as A. It is denoted by A^2. Similarly, we can form A^3, A^4, and in general, A^n for any positive integer n. Thus A^n is the product of A with itself n times.

We can define the **unit** $n \times n$ matrix to be the matrix having diagonal components all equal to 1, and all other components equal to 0. Thus the unit $n \times n$ matrix, denoted by I_n, looks like this:

$$I_n = \begin{pmatrix} 1 & 0 & 0 & \cdots & 0 \\ 0 & 1 & 0 & \cdots & 0 \\ 0 & 0 & 1 & \cdots & 0 \\ \vdots & \vdots & \vdots & & \vdots \\ 0 & 0 & 0 & 1 & 0 \\ 0 & 0 & 0 & \cdots & 1 \end{pmatrix}.$$

We can then define $A^0 = I$ (the unit matrix of the same size as A). Note that for any two integers r, $s \geq 0$ we have the usual relation

$$A^r A^s = A^s A^r = A^{r+s}.$$

For example, in the Markov process described above, we may express the population vector in the $(n + 1)$-th year as

$$X_{n+1} = A^n X_1,$$

where X_1 is the population vector in the first year.

Warning. It is **not always true** that $AB = BA$. For instance, compute AB and BA in the following cases:

$$A = \begin{pmatrix} 3 & 2 \\ 0 & 1 \end{pmatrix} \qquad B = \begin{pmatrix} 2 & -1 \\ 0 & 5 \end{pmatrix}.$$

You will find two different values. This is expressed by saying that multiplication of matrices is not necessarily commutative. Of course, in some *special* cases, we do have $AB = BA$. For instance, powers of A commute, i.e. we have $A^r A^s = A^s A^r$ as already pointed out above.

We now prove other basic properties of multiplication.

Distributive law. *Let A, B, C be matrices. Assume that A, B can be multiplied, and A, C can be multiplied, and B, C can be added. Then A, $B + C$ can be multiplied, and we have*

$$A(B + C) = AB + AC.$$

If x is a number, then

$$A(xB) = x(AB).$$

Proof. Let A_i be the i-th row of A and let B^k, C^k be the k-th column of B and C, respectively.... Then $B^k + C^k$ is the k-th column of $B + C$. By definition, the ik-component of $A(B + C)$ is $A_i \cdot (B^k + C^k)$. Since

$$A_i \cdot (B^k + C^k) = A_i \cdot B^k + A_i \cdot C^k,$$

our first assertion follows. As for the second, observe that the k-th column of xB is xB^k. Since

$$A_i \cdot xB^k = x(A_i \cdot B^k),$$

our second assertion follows.

Associative law. *Let A, B, C be matrices such that A, B can be multiplied and B, C can be multiplied. Then A, BC can be multiplied. So can AB, C, and we have*

$$(AB)C = A(BC).$$

Proof. Let $A = (a_{ij})$ be an $m \times n$ matrix, let $B = (b_{jk})$ be an $n \times r$ matrix, and let $C = (c_{kl})$ be an $r \times s$ matrix. The product AB is an $m \times r$ matrix, whose ik-component is equal to the sum

$$a_{i1}b_{1k} + a_{i2}b_{2k} + \cdots + a_{in}b_{nk}.$$

We shall abbreviate this sum using our $\sum$ notation by writing

$$\sum_{j=1}^{n} a_{ij}b_{jk}.$$

By definition, the il-component of $(AB)C$ is equal to

$$\sum_{k=1}^{r} \left[\sum_{j=1}^{n} a_{ij}b_{jk} \right] c_{kl} = \sum_{k=1}^{r} \left[\sum_{j=1}^{n} a_{ij}b_{jk}c_{kl} \right].$$

The sum on the right can also be described as the sum of all terms

$$\sum a_{ij} b_{jk} c_{kl},$$

where j, k range over all integers $1 \leq j \leq n$ and $1 \leq k \leq r$, respectively.

If we had started with the jl-component of BC and then computed the il-component of $A(BC)$ we would have found exactly the same sum, thereby proving the desired property.

The above properties are very similar to those of multiplication of numbers, except that the commutative law does not hold.

We can also relate multiplication with the transpose:

Let A, B be matrices of a size such that AB is defined. Then

$$^t(AB) = {}^t B \, {}^t A.$$

In other words, the transpose of the product is equal to the product of the transpose in reverse order.

Proof. Let $A = (a_{ij})$ and $B = (b_{jk})$. Then $AB = C = (c_{ik})$ where

$$c_{ik} = a_{i1} b_{1k} + \cdots + a_{in} b_{nk}$$
$$= b_{1k} a_{i1} + \cdots + b_{nk} a_{in}.$$

Let $^t A = (a'_{ji})$, $^t B = (b'_{kj})$, and $^t C = (c'_{ki})$. Then

$$a'_{ji} = a_{ij}, \qquad b'_{kj} = b_{jk}, \qquad c'_{ki} = c_{ik}.$$

Hence we can reread the above relation as

$$c'_{ki} = b'_{k1} a'_{1i} + \cdots + b'_{kn} a'_{ni},$$

which shows that $^t C = {}^t B \, {}^t A$, as desired.

Example. Instead of writing the system of linear equations $AX = B$ in terms of column vectors, we can write it by taking the transpose, which gives

$$^t X \, {}^t A = {}^t B.$$

If X, B are column vectors, then $^t X$, $^t B$ are row vectors. It is occasionally convenient to rewrite the system in this fashion.

Unlike division with non-zero numbers, **we cannot divide by a matrix**, any more than we could divide by a vector (n-tuple). Under certain circumstances, we can define an inverse as follows. We do this only for square matrices. Let A be an $n \times n$ matrix. An **inverse for** A is a matrix B such that

$$AB = BA = I.$$

Since we multiplied A with B on both sides, the only way this can make sense is if B is also an $n \times n$ matrix. Some matrices do not have inverses. However, **if an inverse exists, then there is only one** (we say that **the inverse is unique, or uniquely determined by** A). This is easy to prove. Suppose that B, C are inverses, so we have

$$AB = BA = I \qquad \text{and} \qquad AC = CA = I.$$

Multiply the equation $BA = I$ on the right with C. Then

$$BAC = IC = C$$

and we have assumed that $AC = I$, so $BAC = BI = B$. This proves that $B = C$. In light of this, the inverse is denoted by

$$A^{-1}.$$

Then A^{-1} is the unique matrix such that

$$\boxed{A^{-1}A = I \qquad \text{and} \qquad AA^{-1} = I.}$$

We shall prove later that if A, B are square matrices of the same size such that $AB = I$ then it follows that also

$$BA = I.$$

In other words, if B is a right inverse for A, then it is also a left inverse. You may assume this for the time being. Thus in verifying that a matrix is the inverse of another, you need only do so on one side.

We shall also find later a way of computing the inverse when it exists. It can be a tedious matter.

Let c be a number. Then the matrix

$$cI = \begin{pmatrix} c & 0 & \cdots\cdots & 0 \\ 0 & c & 0 & \cdots & 0 \\ \vdots & \vdots & \ddots & \vdots \\ 0 & \cdots\cdots\cdots & & c \end{pmatrix}$$

having component c on each diagonal entry and 0 otherwise is called a **scalar matrix**. We can also write it as cI, where I is the unit $n \times n$ matrix. Cf. Exercise 6.

As an application of the formula for the transpose of a product, we shall now see that:

The transpose of an inverse is the inverse of the transpose, that is

$$^t(A^{-1}) = (^tA)^{-1}.$$

Proof. Take the transpose of the relation $AA^{-1} = I$. Then by the rule for the transpose of a product, we get

$$^t(A^{-1})^tA = {}^tI = I$$

because I is equal to its own transpose. Similarly, applying the transpose to the relation $A^{-1}A = I$ yields

$$^tA^t(A^{-1}) = {}^tI = I.$$

Hence $^t(A^{-1})$ is an inverse for ^tA, as was to be shown.

In light of this result, it is customary to omit the parentheses, and to write

$$^tA^{-1}$$

for the inverse of the transpose, which we have seen is equal to the transpose of the inverse.

We end this section with an important example of multiplication of matrices.

Example. Rotations. A special type of 2×2 matrix represents rotations. For each number θ, let $R(\theta)$ be the matrix

$$R(\theta) = \begin{pmatrix} \cos \theta & -\sin \theta \\ \sin \theta & \cos \theta \end{pmatrix}.$$

Let $X = \begin{pmatrix} x \\ y \end{pmatrix}$ be a point on the unit circle. We may write its coordinates x, y in the form

$$x = \cos \varphi, \qquad y = \sin \varphi$$

for some number φ. Then we get, by matrix multiplication:

$$R(\theta)\begin{pmatrix} x \\ y \end{pmatrix} = \begin{pmatrix} \cos\theta & -\sin\theta \\ \sin\theta & \cos\theta \end{pmatrix}\begin{pmatrix} \cos\varphi \\ \sin\varphi \end{pmatrix}$$

$$= \begin{pmatrix} \cos(\theta + \varphi) \\ \sin(\theta + \varphi) \end{pmatrix}.$$

This follows from the **addition formula for sine and cosine**, namely

$$\cos(\theta + \varphi) = \cos\theta\cos\varphi - \sin\theta\sin\varphi,$$
$$\sin(\theta + \varphi) = \sin\theta\cos\varphi + \cos\theta\sin\varphi.$$

An arbitrary point in $\mathbf{R}^2$ can be written in the form

$$rX = \begin{pmatrix} r\cos\varphi \\ r\sin\varphi \end{pmatrix},$$

where r is a number ≥ 0. Since

$$R(\theta)rX = rR(\theta)X,$$

we see that multiplication by $R(\theta)$ also has the effect of rotating rX by an angle θ. Thus rotation by an angle θ can be represented by the matrix $R(\theta)$.

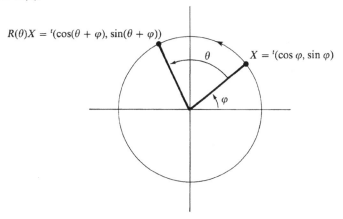

$$R(\theta)X = {}^t(\cos(\theta + \varphi), \sin(\theta + \varphi))$$

$$X = {}^t(\cos\varphi, \sin\varphi)$$

Figure 1

Note that for typographical reasons, we have written the vector ${}^tX$ horizontally, but have put a little t on the upper left superscript, to denote transpose, so X is a column vector.

Example. The matrix corresponding to rotation by an angle of $\pi/3$ is given by

$$R(\pi/3) = \begin{pmatrix} \cos \pi/3 & -\sin \pi/3 \\ \sin \pi/3 & \cos \pi/3 \end{pmatrix}$$

$$= \begin{pmatrix} 1/2 & -\sqrt{3/2} \\ \sqrt{3/2} & 1/2 \end{pmatrix}.$$

Example. Let $X = {}^t(2, 5)$. If you rotate X by an angle of $\pi/3$, find the coordinates of the rotated vector.

These coordinates are:

$$R(\pi/3)X = \begin{pmatrix} 1/2 & -\sqrt{3/2} \\ \sqrt{3/2} & 1/2 \end{pmatrix}\begin{pmatrix} 2 \\ 5 \end{pmatrix}$$

$$= \begin{pmatrix} 1 - 5\sqrt{3/2} \\ \sqrt{3} + 5/2 \end{pmatrix}.$$

Warning. Note how we multiply the column vector on the left with the matrix $R(\theta)$. If you want to work with row vectors, then take the transpose and verify directly that

$$(2, 5)\begin{pmatrix} 1/2 & \sqrt{3/2} \\ -\sqrt{3/2} & 1/2 \end{pmatrix} = (1 - 5\sqrt{3/2}, \sqrt{3} + 5/2).$$

So the matrix $R(\theta)$ gets transposed. The minus sign is now in the lower left-hand corner.

Exercises II, §2

The following exercises give mostly routine practice in the multiplication of matrices. However, they also illustrate some more theoretical aspects of this multiplication. Therefore they should be all worked out. Specifically:

Exercises 7 through 12 illustrate multiplication by the standard unit vectors.

Exercises 14 through 19 illustrate multiplication of triangular matrices.

Exercises 24 through 27 illustrate how addition of numbers is transformed into multiplication of matrices.

Exercises 27 through 32 illustrate rotations.

Exercises 33 through 37 illustrate elementary matrices, *and should be worked out before studying §5.*

1. Let I be the unit $n \times n$ matrix. Let A be an $n \times r$ matrix. What is IA? If A is an $m \times n$ matrix, what is AI?

2. Let O be the matrix all of whose coordinates are 0. Let A be a matrix of a size such that the product AO is defined. What is AO?

3. In each one of the following cases, find $(AB)C$ and $A(BC)$.

(a) $A = \begin{pmatrix} 2 & 1 \\ 3 & 1 \end{pmatrix}$, $B = \begin{pmatrix} -1 & 1 \\ 1 & 0 \end{pmatrix}$, $C = \begin{pmatrix} 1 & 4 \\ 2 & 3 \end{pmatrix}$

(b) $A = \begin{pmatrix} 2 & 1 & -1 \\ 3 & 1 & 2 \end{pmatrix}$, $B = \begin{pmatrix} 1 & 1 \\ 2 & 0 \\ 3 & -1 \end{pmatrix}$, $C = \begin{pmatrix} 1 \\ 3 \end{pmatrix}$

(c) $A = \begin{pmatrix} 2 & 4 & 1 \\ 3 & 0 & -1 \end{pmatrix}$, $B = \begin{pmatrix} 1 & 1 & 0 \\ 2 & 1 & -1 \\ 3 & 1 & 5 \end{pmatrix}$, $C = \begin{pmatrix} 1 & 2 \\ 3 & 1 \\ -1 & 4 \end{pmatrix}$

4. Let A, B be square matrices of the same size, and assume that $AB = BA$. Show that

$$(A + B)^2 = A^2 + 2AB + B^2, \qquad \text{and} \qquad (A + B)(A - B) = A^2 - B^2,$$

using the distributive law.

5. Let

$$A = \begin{pmatrix} 1 & 2 \\ 3 & -1 \end{pmatrix}, \qquad B = \begin{pmatrix} 2 & 0 \\ 1 & 1 \end{pmatrix}.$$

Find AB and BA.

6. Let

$$C = \begin{pmatrix} 7 & 0 \\ 0 & 7 \end{pmatrix}.$$

Let A, B be as in Exercise 5. Find CA, AC, CB, and BC. State the general rule including this exercise as a special case.

7. Let $X = (1, 0, 0)$ and let

$$A = \begin{pmatrix} 3 & 1 & 5 \\ 2 & 0 & 1 \\ 1 & 1 & 7 \end{pmatrix}.$$

What is XA?

8. Let $X = (0, 1, 0)$, and let A be an arbitrary 3×3 matrix. How would you describe XA? What if $X = (0, 0, 1)$? Generalize to similar statements concerning $n \times n$ matrices, and their products with unit vectors.

9. Let

$$A = \begin{pmatrix} 2 & 1 & 3 \\ 4 & 1 & 5 \end{pmatrix}.$$

Find AX for each of the following values of X.

(a) $X = \begin{pmatrix} 1 \\ 0 \\ 0 \end{pmatrix}$ (b) $X = \begin{pmatrix} 0 \\ 1 \\ 1 \end{pmatrix}$ (c) $X = \begin{pmatrix} 0 \\ 0 \\ 1 \end{pmatrix}$

10. Let

$$A = \begin{pmatrix} 3 & 7 & 5 \\ 1 & -1 & 4 \\ 2 & 1 & 8 \end{pmatrix}.$$

Find AX for each of the values of X given in Exercise 9.

11. Let

$$X = \begin{pmatrix} 0 \\ 1 \\ 0 \\ 0 \end{pmatrix} \quad \text{and} \quad A = \begin{pmatrix} a_{11} & \cdots & a_{14} \\ \vdots & & \vdots \\ a_{m1} & \cdots & a_{m4} \end{pmatrix}.$$

What is AX?

12. Let X be a column vector having all its components equal to 0 except the j-th component which is equal to 1. Let A be an arbitrary matrix, whose size is such that we can form the product AX. What is AX?

13. Let X be the indicated column vector, and A the indicated matrix. Find AX as a column vector.

(a) $X = \begin{pmatrix} 3 \\ 2 \\ 1 \end{pmatrix}$, $A = \begin{pmatrix} 1 & 0 & 1 \\ 2 & 1 & 1 \\ 2 & 0 & -1 \end{pmatrix}$ (b) $X = \begin{pmatrix} 1 \\ 1 \\ 0 \end{pmatrix}$, $A = \begin{pmatrix} 2 & 1 & 5 \\ 0 & 1 & 1 \end{pmatrix}$

(c) $X = \begin{pmatrix} x_1 \\ x_2 \\ x_3 \end{pmatrix}$, $A = \begin{pmatrix} 0 & 1 & 0 \\ 0 & 0 & 0 \end{pmatrix}$ (d) $X = \begin{pmatrix} x_1 \\ x_2 \\ x_3 \end{pmatrix}$, $A = \begin{pmatrix} 0 & 0 & 0 \\ 1 & 0 & 0 \end{pmatrix}$

14. Let $A = \begin{pmatrix} a & b \\ c & d \end{pmatrix}$. Find the product AS for each one of the following matrices S. Describe in words the effect on A of this product.

(a) $S = \begin{pmatrix} 1 & x \\ 0 & 1 \end{pmatrix}$ (b) $S = \begin{pmatrix} 1 & 0 \\ x & 1 \end{pmatrix}$.

15. Let $A = \begin{pmatrix} a & b \\ c & d \end{pmatrix}$ again. Find the product SA for each one of the following matrices S. Describe in words the effect of this product on A.

(a) $S = \begin{pmatrix} 1 & x \\ 0 & 1 \end{pmatrix}$ (b) $S = \begin{pmatrix} 1 & 0 \\ x & 1 \end{pmatrix}$.

16. (a) Let A be the matrix

$$\begin{pmatrix} 0 & 1 & 1 \\ 0 & 0 & 1 \\ 0 & 0 & 0 \end{pmatrix}.$$

Find A^2, A^3. Generalize to 4×4 matrices.

(b) Let A be the matrix

$$\begin{pmatrix} 1 & 1 & 1 \\ 0 & 1 & 1 \\ 0 & 0 & 1 \end{pmatrix}.$$

Compute A^2, A^3, A^4.

17. Let

$$A = \begin{pmatrix} 1 & 0 & 0 \\ 0 & 2 & 0 \\ 0 & 0 & 3 \end{pmatrix}.$$

Find A^2, A^3, A^4.

18. Let A be a diagonal matrix, with diagonal elements $a_1, \ldots, a_n$. What is A^2, A^3, A^k for any positive integer k?

19. Let

$$A = \begin{pmatrix} 0 & 1 & 6 \\ 0 & 0 & 4 \\ 0 & 0 & 0 \end{pmatrix}.$$

Find A^3

20. (a) Find a 2×2 matrix A such that $A^2 = -I = \begin{pmatrix} -1 & 0 \\ 0 & -1 \end{pmatrix}$.

(b) Determine all 2×2 matrices A such that $A^2 = O$.

21. Let A be a square matrix.

(a) If $A^2 = O$ show that $I - A$ is invertible.

(b) If $A^3 = O$, show that $I - A$ is invertible.

(c) In general, if $A^n = O$ for some positive integer n, show that $I - A$ is invertible. [*Hint*: Think of the geometric series.]

(d) Suppose that $A^2 + 2A + I = O$. Show that A is invertible.

(e) Suppose that $A^3 - A + I = O$. Show that A is invertible.

22. Let A, B be two square matrices of the same size. We say that A is **similar** to B if there exists an invertible matrix T such that $B = TAT^{-1}$. Suppose this is the case. Prove:

(a) B is similar to A.

(b) A is invertible if and only if B is invertible.

(c) ^tA is similar to ^tB.

(d) Suppose $A^n = O$ and B is an invertible matrix of the same size as A. Show that $(BAB^{-1})^n = O$.

23. Let A be a square matrix which is of the form

$$\begin{pmatrix} a_{11} & * & \cdots\cdots & * \\ 0 & a_{22} & * & \cdots & * \\ \vdots & & \ddots & \vdots \\ & & & * \\ 0 & \cdots\cdots & 0 & a_{nn} \end{pmatrix}.$$

The notation means that all elements below the diagonal are equal to 0, and the elements above the diagonal are arbitrary. One may express this property by saying that

$$a_{ij} = 0 \quad \text{if} \quad i > j.$$

Such a matrix is called **upper triangular**. If A, B are upper triangular matrices (of the same size) what can you say about the diagonal elements of AB?

Exercises 24 through 27 give examples where addition of numbers is transformed into multiplication of matrices.

24. Let a, b be numbers, and let

$$A = \begin{pmatrix} 1 & a \\ 0 & 1 \end{pmatrix} \quad \text{and} \quad B = \begin{pmatrix} 1 & b \\ 0 & 1 \end{pmatrix}.$$

What is AB? What is A^2, A^3? What is A^n where n is a positive integer?

25. Show that the matrix A in Exercise 24 has an inverse. What is this inverse?

26. Show that if A, B are $n \times n$ matrices which have inverses, then AB has an inverse.

27. **Rotations.** Let $R(\theta)$ be the matrix given by

$$R(\theta) = \begin{pmatrix} \cos \theta & -\sin \theta \\ \sin \theta & \cos \theta \end{pmatrix}.$$

(a) Show that for any two numbers θ_1, θ_2 we have

$$R(\theta_1)R(\theta_2) = R(\theta_1 + \theta_2).$$

[You will have to use the addition formulas for sine and cosine.]
(b) Show that the matrix $R(\theta)$ has an inverse, and write down this inverse.
(c) Let $A = R(\theta)$. Show that

$$A^2 = \begin{pmatrix} \cos 2\theta & -\sin 2\theta \\ \sin 2\theta & \cos 2\theta \end{pmatrix}.$$

(d) Determine A^n for any positive integer n. Use induction.

28. Find the matrix $R(\theta)$ associated with the rotation for each of the following values of θ.
(a) $\pi/2$ (b) $\pi/4$ (c) π (d) $-\pi$ (e) $-\pi/3$
(f) $\pi/6$ (g) $5\pi/4$

29. In general, let $\theta > 0$. What is the matrix associated with the rotation by an angle $-\theta$ (i.e. clockwise rotation by θ)?

30. Let $X = {}^t(1, 2)$ be a point of the plane. If you rotate X by an angle of $\pi/4$, what are the coordinates of the new point?

31. Same question when $X = {}^t(-1, 3)$ and the rotation is by an angle of $\pi/2$.

32. For any vector X in $\mathbf{R}^2$ let $Y = R(\theta)X$ be its rotation by an angle θ. Show that $\|Y\| = \|X\|$.

The following exercises on elementary matrices should be done before studying §5.

33. **Elementary matrices.** Let

$$A = \begin{pmatrix} 2 & 3 & -1 & 1 \\ 1 & 4 & 2 & -2 \\ -1 & 1 & 3 & -5 \\ 1 & 2 & 3 & 4 \end{pmatrix}$$

Let U be the matrix as shown. In each case find UA.

(a) $\begin{pmatrix} 0 & 1 & 0 & 0 \\ 0 & 0 & 0 & 0 \\ 0 & 0 & 0 & 0 \\ 0 & 0 & 0 & 0 \end{pmatrix}$
(b) $\begin{pmatrix} 0 & 0 & 0 & 0 \\ 1 & 0 & 0 & 0 \\ 0 & 0 & 0 & 0 \\ 0 & 0 & 0 & 0 \end{pmatrix}$

(c) $\begin{pmatrix} 0 & 0 & 0 & 0 \\ 0 & 0 & 0 & 0 \\ 0 & 1 & 0 & 0 \\ 0 & 0 & 0 & 0 \end{pmatrix}$
(d) $\begin{pmatrix} 0 & 0 & 0 & 0 \\ 0 & 0 & 1 & 0 \\ 0 & 0 & 0 & 0 \\ 0 & 0 & 0 & 0 \end{pmatrix}$

(e) $\begin{pmatrix} 0 & 0 & 0 & 0 \\ 0 & 0 & 0 & 0 \\ 0 & 0 & 0 & 0 \\ 0 & 0 & 1 & 0 \end{pmatrix}$
(f) $\begin{pmatrix} 0 & 0 & 0 & 0 \\ 0 & 0 & 0 & 0 \\ 0 & 0 & 0 & 1 \\ 0 & 0 & 0 & 0 \end{pmatrix}$

34. Let E be the matrix as shown. Find EA where A is the same matrix as in the preceding exercise.

(a) $\begin{pmatrix} 0 & 1 & 0 & 0 \\ 1 & 0 & 0 & 0 \\ 0 & 0 & 1 & 0 \\ 0 & 0 & 0 & 1 \end{pmatrix}$
(b) $\begin{pmatrix} 1 & 0 & 0 & 0 \\ 0 & 0 & 1 & 0 \\ 0 & 1 & 0 & 0 \\ 0 & 0 & 0 & 1 \end{pmatrix}$

(c) $\begin{pmatrix} 1 & 0 & 0 & 0 \\ 0 & 1 & 0 & 0 \\ 0 & 0 & 1 & 0 \\ 0 & 5 & 0 & 1 \end{pmatrix}$
(d) $\begin{pmatrix} 1 & 0 & 0 & 0 \\ 0 & 1 & 0 & 0 \\ 0 & -2 & 1 & 0 \\ 0 & 0 & 0 & 1 \end{pmatrix}$

35. Let E be the matrix as shown. Find EA where A is the same matrix as in the preceding exercise and Exercise 33.

$$(a) \begin{pmatrix} 3 & 0 & 0 & 0 \\ 0 & 1 & 0 & 0 \\ 0 & 0 & 1 & 0 \\ 0 & 0 & 0 & 1 \end{pmatrix} \qquad (b) \begin{pmatrix} 1 & 0 & 3 & 0 \\ 0 & 1 & 0 & 0 \\ 0 & 0 & 1 & 0 \\ 0 & 0 & 0 & 1 \end{pmatrix}$$

$$(c) \begin{pmatrix} 1 & 0 & 0 & 0 \\ -2 & 1 & 0 & 0 \\ 0 & 0 & 1 & 0 \\ 0 & 0 & 0 & 1 \end{pmatrix} \qquad (d) \begin{pmatrix} 1 & 0 & 0 & 0 \\ 0 & 1 & 0 & 0 \\ 0 & -2 & 1 & 0 \\ 0 & 0 & 0 & 1 \end{pmatrix}$$

36. Let $A = (a_{ij})$ be an $m \times n$ matrix,

$$\begin{pmatrix} a_{11} & \cdots & a_{1n} \\ \vdots & & \vdots \\ a_{m1} & \cdots & a_{mn} \end{pmatrix}.$$

Let $1 \leq r \leq m$ and $1 \leq s \leq m$. Let I_{rs} be the matrix whose rs-component is 1 and such that all other components are equal to 0.
(a) What is $I_{rs}A$?
(b) Suppose $r \neq s$. What is $(I_{rs} + I_{sr})A$?
(c) Suppose $r \neq s$. Let I_{jj} be the matrix whose jj-component is 1 and such that all other components are 0. Let

$$E_{rs} = I_{rs} + I_{sr} + \text{sum of all } I_{jj} \text{ for } \quad j \neq r, \quad j \neq s.$$

What is $E_{rs}A$?

37. Again let $r \neq s$.
(a) Let $E = I + 3I_{rs}$. What is EA?
(b) Let c be any number. Let $E = I + cI_{rs}$. What is EA?

The rest of the chapter will be mostly concerned with linear equations, and especially homogeneous ones. We shall find three ways of interpreting such equations, illustrating three dfferent ways of thinking about matrices and vectors.

II, §3. Homogeneous Linear Equations and Elimination

In this section, we look at linear equations by one method of elimination. In the next section, we shall discuss another method.

We shall be interested in the case when the number of unknowns is greater than the number of equations, and we shall see that in that case, there always exists a non-trivial solution.

Before dealing with the general case, we shall study examples.

Example 1. Suppose that we have a single equation, like

$$2x + y - 4z = 0.$$

We wish to find a solution with not all of x, y, z equal to 0. An equivalent equation is

$$2x = -y + 4z.$$

To find a non-trivial solution, we give all the variables except the first a special value $\neq 0$, say $y = 1$, $z = 1$. We than solve for x. We find

$$2x = -y + 4z = 3,$$

whence $x = \frac{3}{2}$.

Example 2. Consider a pair of equations, say

(1) $$2x + 3y - z = 0,$$

(2) $$x + y + z = 0.$$

We reduce the problem of solving these simultaneous equations to the preceding case of one equation, by eliminating one variable. Thus we multiply the second equation by 2 and subtract it from the first equation, getting

(3) $$y - 3z = 0.$$

Now we meet one equation in more than one variable. We give z any value $\neq 0$, say $z = 1$, and solve for y, namely $y = 3$. We then solve for x from the second equation, namely $x = -y - z$, and obtain $x = -4$. The values which we have obtained for x, y, z are also solutions of the first equation, because the first equation is (in an obvious sense) the sum of equation (2) multiplied by 2, and equation (3).

Example 3. We wish to find a solution for the system of equations

$$3x - 2y + z + 2w = 0,$$
$$x + y - z - w = 0,$$
$$2x - 2y + 3z = 0.$$

Again we use the elimination method. Multiply the second equation by 2 and subtract it from the third. We find

$$-4y + 5z + 2w = 0.$$

Multiply the second equation by 3 and subtract it from the first. We find

$$-5y + 4z + 5w = 0.$$

We have now eliminated x from our equations, and find two equations in three unknowns, y, z, w. We eliminate y from these two equations as follows: Multiply the top one by 5, multiply the bottom one by 4, and subtract them. We get

$$9z - 10w = 0.$$

Now give an arbitrary value $\neq 0$ to w, say $w = 1$. Then we can solve for z, namely

$$z = 10/9.$$

Going back to the equations before that, we solve for y, using

$$4y = 5z + 2w.$$

This yields

$$y = 17/9.$$

Finally we solve for x using say the second of the original set of three equations, so that

$$x = -y + z + w,$$

or numerically,

$$x = -49/9.$$

Thus we have found:

$$w = 1, \qquad z = 10/9, \qquad y = 68/9, \qquad x = -49/9.$$

Note that we had three equations in four unknowns. By a successive elimination of variables, we reduced these equations to two equations in three unknowns, and then one equation in two unknowns.

Using precisely the same method, suppose that we start with three equations in five unknowns. Eliminating one variable will yield two equations in four unknowns. Eliminating another variable will yield one equation in three unknowns. We can then solve this equation, and proceed backwards to get values for the previous variables just as we have shown in the examples.

In general, suppose that we start with m equations with n unknowns, and $n > m$. We eliminate one of the variables, say x_1, and obtain a system of $m - 1$ equations in $n - 1$ unknowns. We eliminate a second variable, say x_2, and obtain a system of $m - 2$ equations in $n - 2$ unknowns. Proceeding stepwise, we eliminate $m - 1$ variables, ending up with 1 equation in $n - m + 1$ unknowns. We then give non-trivial arbitrary values to all the remaining variables but one, solve for this last variable, and then proceed backwards to solve successively for each one of the eliminated variables as we did in our examples. Thus we have an effective way of finding a non-trivial solution for the original system.

We shall phrase this in terms of induction in a precise manner.

Let $A = (a_{ij})$, $i = 1, \ldots, m$ and $j = 1, \ldots, n$ be a matrix. Let $b_1, \ldots, b_m$ be numbers. Equations like

$$
\begin{aligned}
a_{11} x_1 + \cdots + a_{1n} x_n &= b_1 \\
&\vdots \\
a_{m1} x_1 + \cdots + a_{mn} x_n &= b_m
\end{aligned}
$$

(*)

are called linear equations. We also say that (*) is a system of linear equations. The system is said to be **homogeneous** if all the numbers $b_1, \ldots, b_m$ are equal to 0. The number n is called the number of **unknowns**, and m is the number of equations.

The system of equations

$$
\begin{aligned}
a_{11} x_1 + \ldots + a_{1n} x_n &= 0 \\
&\vdots \\
a_{m1} x_1 + \ldots + a_{mn} x_n &= 0
\end{aligned}
$$

(**)

will be called the **homogeneous system associated with** (*). In this section, we study the homogeneous system (**).

The system (**) always has a solution, namely the solution obtained by letting all $x_i = 0$. This solution will be called the **trivial** solution. A solution $(x_1, \ldots, x_n)$ such that some x_i is $\neq 0$ is called **non-trivial**.

Consider our system of homogeneous equations (**). Let $A_1, \ldots, A_m$ be the row vectors of the matrix (a_{ij}). Then we can rewrite our equations (**) in the form

$$
\begin{aligned}
A_1 \cdot X &= 0 \\
&\vdots \\
A_m \cdot X &= 0.
\end{aligned}
$$

(**)

Therefore a solution of the system of linear equations can be interpreted as the set of all n-tuples X which are perpendicular to the row vectors of the matrix A. Geometrically, to find a solution of (**) amounts to finding a vector X which is perpendicular to $A_1, \ldots, A_m$. Using the notation of the dot product will make it easier to formulate the proof of our main theorem, namely:

Theorem 3.1. *Let*

(**)
$$a_{11}x_1 + \cdots + a_{1n}x_n = 0$$
$$\vdots \qquad\qquad \vdots$$
$$a_{m1}x_1 + \cdots + a_{mn}x_n = 0$$

be a system of m linear equations in n unknowns, and assume that $n > m$. Then the system has a non-trivial solution.

Proof. The proof will be carried out by induction.
Consider first the case of one equation in n unknowns, $n > 1$:

$$a_1x_1 + \cdots + a_nx_n = 0.$$

If all coefficients $a_1, \ldots, a_n$ are equal to 0, then any value of the variables will be a solution, and a non-trivial solution certainly exists. Suppose that some coefficient a_i is $\neq 0$. After renumbering the variables and the coefficients, we may assume that it is a_1. Then we give $x_2, \ldots, x_n$ arbitrary values, for instance we let $x_2 = \cdots = x_n = 1$, and solve for x_1, letting

$$x_1 = \frac{-1}{a_1}(a_2 + \cdots + a_n).$$

In that manner, we obtain a non-trivial solution for our system of equations.

Let us now assume that our theorem is true for a system of $m - 1$ equations in more than $m - 1$ unknowns. We shall prove that it is true for m equations in n unknowns when $n > m$. We consider the system (**).

If all coefficients (a_{ij}) are equal to 0, we can give any non-zero value to our variables to get a solution. If some coefficient is not equal to 0, then after renumbering the equations and the variables, we may assume that it is a_{11}. We shall subtract a multiple of the first equation from the others to eliminate x_1. Namely, we consider the system of equations

$$\left(A_2 - \frac{a_{21}}{a_{11}}A_1\right) \cdot X = 0$$
$$\vdots$$
$$\left(A_m - \frac{a_{m1}}{a_{11}}A_1\right) \cdot X = 0,$$

which can also be written in the form

(***)
$$A_2 \cdot X - \frac{a_{21}}{a_{11}}A_1 \cdot X = 0$$
$$\vdots$$
$$A_m \cdot X - \frac{a_{m1}}{a_{11}}A_1 \cdot X = 0.$$

In this system, the coefficient of x_1 is equal to 0. Hence we may view
(***) as a system of $m - 1$ equations in $n - 1$ unknowns, and we have
$n - 1 > m - 1$.

According to our assumption, we can find a non-trivial solution
$(x_2, \ldots, x_n)$ for this system. We can then solve for x_1 in the first equation, namely

$$x_1 = \frac{-1}{a_{11}} (a_{12}x_2 + \cdots + a_{1n}x_n).$$

In that way, we find a solution of $A_1 \cdot X = 0$. But according to (***), we
have

$$A_i \cdot X = \frac{a_{i1}}{a_{11}} A_1 \cdot X$$

for $i = 2, \ldots, m$. Hence $A_i \cdot X = 0$ for $i = 2, \ldots, m$, and therefore we have
found a non-trivial solution to our original system (**).

The argument we have just given allows us to proceed stepwise from
one equation to two equations, then from two to three, and so forth.
This concludes the proof.

Exercises II, §3

1. Let

$$E_1 = (1, 0, \ldots, 0), \quad E_2 = (0, 1, 0, \ldots, 0), \quad \ldots, \quad E_n = (0, \ldots, 0, 1)$$

be the standard unit vectors of $\mathbf{R}^n$. Let X be an n-tuple. If $X \cdot E_i = 0$ for all i,
show that $X = O$.

2. Let $A_1, \ldots, A_m$ be vectors in $\mathbf{R}^n$. Let X, Y be solutions of the system of equations

$$X \cdot A_i = 0 \quad \text{and} \quad Y \cdot A_i = 0 \quad \text{for} \quad i = 1, \ldots, m.$$

Show that $X + Y$ is also a solution. If c is a number, show that cX is a
solution.

3. In Exercise 2, suppose that X is perpendicular to each one of the vectors
$A_1, \ldots, A_m$. Let $c_1, \ldots, c_m$ be numbers. A vector

$$c_1 A_1 + \cdots + c_m A_m$$

is called a **linear combination** of $A_1, \ldots, A_m$. Show that X is perpendicular to
such a vector.

4. Consider the inhomogeneous system (∗) consisting of all X such that $X \cdot A_i = b_i$ for $i = 1, \ldots, m$. If X and X' are two solutions of this system, show that there exists a solution Y of the homogeneous system (∗∗) such that $X' = X + Y$. Conversely, if X is any solution of (∗), and Y a solution of (∗∗), show that $X + Y$ is a solution of (∗).

5. Find at least one non-trivial solution for each one of the following systems of equations. Since there are many choices involved, we don't give answers.

(a) $3x + y + z = 0$

(b) $3x + y + z = 0$
$x + y + z = 0$

(c) $2x - 3y + 4z = 0$
$3x + y + z = 0$

(d) $\quad 2x + y + 4z + w = 0$
$-3x + 2y - 3z + w = 0$
$x + y + z = 0$

(e) $-x + 2y - 4z + w = 0$
$x + 3y + z - w = 0$

(f) $-2x + 3y + z + 4w = 0$
$x + y + 2z + 3w = 0$
$2x + y + z - 2w = 0$

6. Show that the only solutions of the following systems of equations are trivial.

(a) $2x + 3y = 0$
$x - y = 0$

(b) $\quad 4x + 5y = 0$
$-6x + 7y = 0$

(c) $\quad 3x + 4y - 2z = 0$
$x + y + z = 0$
$-x - 3y + 5z = 0$

(d) $4x - 7y + 3z = 0$
$x + y = 0$
$y - 6z = 0$

(e) $7x - 2y + 5z + w = 0$
$x - y + z = 0$
$y - 2z + w = 0$
$x + z + w = 0$

(f) $\quad -3x + y + z = 0$
$x - y + z - 2w = 0$
$x - z + w = 0$
$-x + y - 3w = 0$

II, §4. Row Operations and Gauss Elimination

Consider the system of linear equations

$$3x - 2y + \quad z + 2w = 1,$$
$$x + \quad y - \quad z - \quad w = -2,$$
$$2x - \quad y + 3z \qquad = 4.$$

The matrix of coefficients is

$$\begin{pmatrix} 3 & -2 & 1 & 2 \\ 1 & 1 & -1 & -1 \\ 2 & -1 & 3 & 0 \end{pmatrix}.$$

By the **augmented matrix** we shall mean the matrix obtained by inserting the column

$$\begin{pmatrix} 1 \\ -2 \\ 4 \end{pmatrix}$$

as a last column, so the augmented matrix is

$$\begin{pmatrix} 3 & -2 & 1 & 2 & 1 \\ 1 & 1 & -1 & -1 & -2 \\ 2 & -1 & 3 & 0 & 4 \end{pmatrix}.$$

In general, let $AX = B$ be a system of m linear equations in n unknowns, which we write in full:

$$a_{11}x_1 + \cdots + a_{1n}x_n = b_1,$$
$$a_{21}x_1 + \cdots + a_{2n}x_n = b_2,$$
$$\vdots \qquad \qquad \vdots \quad \vdots$$
$$a_{m1}x_1 + \cdots + a_{mn}x_n = b_m.$$

Then we define the **augmented matrix** to be the m by $n + 1$ matrix:

$$\begin{pmatrix} a_{11} & a_{12} & \cdots & a_{1n} & b_1 \\ a_{21} & a_{22} & \cdots & a_{2n} & b_2 \\ \vdots & & & \vdots & \vdots \\ a_{m1} & a_{m2} & \cdots & a_{mn} & b_m \end{pmatrix}.$$

In the examples of homogeneous linear equations of the preceding section, you will notice that we performed the following operations, called **elementary row operations**:

Multiply one equation by a non-zero number.
Add one equation to another.
Interchange two equations.

These operations are reflected in operations on the augmented matrix of coefficients, which are also called **elementary row operations**:

Multiply one row by a non-zero number.
Add one row to another.
Interchange two rows.

Suppose that a system of linear equations is changed by an elementary row operation. Then the solutions of the new system are exactly the same as the solutions of the old system. By making row operations, we can hope to simplify the shape of the system so that it is easier to find the solutions.

Let us define two matrices to be **row equivalent** if one can be obtained from the other by a succession of elementary row operations. If A is the matrix of coefficients of a system of linear equations, and B the column vector as above, so that

$$(A, B)$$

is the augmented matrix, and if (A', B') is row-equivalent to (A, B) then the solutions of the system

$$AX = B$$

are the same as the solutions of the system

$$A'X = B'.$$

To obtain an equivalent system (A', B') as simple as possible we use a method which we first illustrate in a concrete case.

Example. Consider the augmented matrix in the above example. We have the following row equivalences:

$$\begin{pmatrix} 3 & -2 & 1 & 2 & 1 \\ 1 & 1 & -1 & -1 & -2 \\ 2 & -1 & 3 & 0 & 4 \end{pmatrix}$$

Subtract 3 times second row from first row

$$\begin{pmatrix} 0 & -5 & 4 & 5 & 7 \\ 1 & 1 & -1 & -1 & -2 \\ 2 & -1 & 3 & 0 & 4 \end{pmatrix}$$

Subtract 2 times second row from third row

$$\begin{pmatrix} 0 & -5 & 4 & 5 & 7 \\ 1 & 1 & -1 & -1 & -2 \\ 0 & -3 & 5 & 2 & 8 \end{pmatrix}$$

Interchange first and second row; multiply second row by -1.

$$\begin{pmatrix} 1 & 1 & -1 & -1 & -2 \\ 0 & 5 & -4 & -5 & -7 \\ 0 & -3 & 5 & 2 & 8 \end{pmatrix}$$

Multiply second row by 3; multiply third row by 5.

$$\begin{pmatrix} 1 & 1 & -1 & -1 & -2 \\ 0 & 15 & -12 & -15 & -21 \\ 0 & -15 & 25 & 10 & 40 \end{pmatrix}$$

Add second row to third row.

$$\begin{pmatrix} 1 & 1 & -1 & -1 & -2 \\ 0 & 15 & -12 & -15 & -21 \\ 0 & 0 & 13 & -5 & 19 \end{pmatrix}$$

What we have achieved is to make each successive row start with a non-zero entry at least one step further than the preceding row. This makes it very simple to solve the equations. The new system whose augmented matrix is the matrix obtained last can be written in the form:

$$x + y - z - w = -2,$$

$$15y - 12z - 15w = -21,$$

$$13z - 5w = 19.$$

This is now in a form where we can solve by giving w an arbitrary value in the third equation, and solve for z from the third equation. Then we solve for y from the second, and x from the first. With the formulas, this gives:

$$z = \frac{19 + 5w}{13},$$

$$y = \frac{-21 + 12z + 15w}{15},$$

$$x = -1 - y + z + w.$$

We can give w any value to start with, and then determine values for x, y, z. Thus we see that the solutions depend on one free parameter. Later we shall express this property by saying that the set of solutions has dimension 1.

For the moment, we give a general name to the above procedure. Let M be a matrix. We shall say that M is in **row echelon form** if it has the following property:

Whenever two successive rows do not consist entirely of zeros, then the second row starts with a non-zero entry at least one step further to the right than the first row. All the rows consisting entirely of zeros are at the bottom of the matrix.

In the previous example we transformed a matrix into another which is in row echelon form. The non-zero coefficients occurring furthest to

the left in each row are called the **leading coefficients**. In the above
example, the leading coefficients are 1, 15, 13. One may perform one
more change by dividing each row by the leading coefficient. Then the
above matrix is row equivalent to

$$\begin{pmatrix} 1 & 1 & -1 & -1 & -1 \\ 0 & 1 & -\frac{4}{5} & -1 & -\frac{7}{5} \\ 0 & 0 & 1 & -\frac{5}{13} & \frac{19}{13} \end{pmatrix}.$$

In this last matrix, the leading coefficient of each row is equal to 1. One
could make further row operations to insert further zeros, for instance
subtract the second row from the first, and then subtract $\frac{2}{5}$ times the
third row from the second. This yields:

$$\begin{pmatrix} 1 & 0 & -\frac{7}{5} & -\frac{6}{5} & -2 \\ 0 & 1 & 0 & \frac{1}{5}+\frac{2}{13} & 1-\frac{38}{65} \\ 0 & 0 & 1 & -\frac{5}{13} & \frac{19}{13} \end{pmatrix}.$$

Unless the matrix is rigged so that the fractions do not look too horrible, it is usually a pain to do this further row equivalence by hand, but
a machine would not care.

   Example. The following matrix is in row echelon form.

$$\begin{pmatrix} 0 & 2 & -3 & 4 & 1 & 7 \\ 0 & 0 & 0 & 5 & 2 & -4 \\ 0 & 0 & 0 & 0 & -3 & 1 \\ 0 & 0 & 0 & 0 & 0 & 0 \end{pmatrix}.$$

Suppose that this matrix is the augmented matrix of a system of linear
equations, then we can solve the linear equations by giving some variables an arbitrary value as we did. Indeed, the equations are:

$$2y - 3z + 4w + \ t = 7,$$

$$5w + 2t = -4,$$

$$-3t = 1.$$

Then the solutions are

$$t = -1/3,$$

$$w = \frac{-4 - 2t}{5},$$

$$z = \text{any arbitrarily given value},$$

$$y = \frac{7 + 3z - 4w - t}{2},$$

$$x = \text{any arbitrarily given value}.$$

The method of changing a matrix by row equivalences to put it in row echelon form works in general.

Theorem 4.1. *Every matrix is row equivalent to a matrix in row echelon form.*

Proof. Select a non-zero entry furthest to the left in the matrix. If this entry is not in the first column, this means that the matrix consists entirely of zeros to the left of this entry, and we can forget about them. So suppose this non-zero entry is in the first column. After an interchange of rows, we can find an equivalent matrix such that the upper left-hand corner is not 0. Say the matrix is

$$\begin{pmatrix} a_{11} & a_{12} & \cdots & a_{1n} \\ a_{21} & a_{22} & \cdots & a_{2n} \\ \vdots & \vdots & & \vdots \\ a_{m1} & a_{m2} & \cdots & a_{mn} \end{pmatrix}$$

and $a_{11} \neq 0$. We multiply the first row by a_{21}/a_{11} and subtract from the second row. Similarly, we multiply the first row by a_{i1}/a_{11} and subtract it from the i-th row. Then we obtain a matrix which has zeros in the first column except for a_{11}. Thus the original matrix is row equivalent to a matrix of the form

$$\begin{pmatrix} a_{11} & a_{12} & \cdots & a_{1n} \\ 0 & a'_{22} & \cdots & a'_{2n} \\ \vdots & \vdots & & \vdots \\ 0 & a'_{m2} & \cdots & a'_{mn} \end{pmatrix}.$$

We then repeat the procedure with the smaller matrix

$$\begin{pmatrix} a'_{22} & \cdots & a'_{2n} \\ \vdots & & \vdots \\ a'_{m2} & \cdots & a'_{mn} \end{pmatrix}.$$

We can continue until the matrix is in row echelon form (formally by induction). This concludes the proof.

Observe that the proof is just another way of formulating the elimination argument of §3.

We give another proof of the fundamental theorem:

Theorem 4.2. *Let*

$$a_{11}x_1 + \cdots + a_{1n}x_n = 0,$$
$$\vdots \qquad\qquad \vdots$$
$$a_{m1}x_1 + \cdots + a_{mn}x_n = 0,$$

*be a system of m homogeneous linear equations in n unknowns with
n > m. Then there exists a non-trivial solution.*

Proof. Let $A = (a_{ij})$ be the matrix of coefficients. Then A is equivalent to A' in row echelon form:

$$a_{k_1}x_{k_1} + S_{k_1}(x) \qquad\qquad = 0,$$
$$a_{k_2}x_{k_2} + S_{k_2}(x) \qquad\quad = 0,$$
$$\cdots\cdots\cdots\cdots$$
$$a_{k_r}x_{k_r} + S_{k_r}(x) = 0,$$

where $a_{k_1} \neq 0, \ldots, a_{k_r} \neq 0$ are the non-zero coefficients of the variables
occurring furthest to the left in each successive row, and $S_{k_1}(x), \ldots, S_{k_r}(x)$
indicate sums of variables with certain coefficients, but such that if a
variable x_j occurs in $S_{k_1}(x)$, then $j > k_1$ and similarly for the other sums.
If x_j occurs in S_{k_i} then $j > k_i$. Since by assumption the total number of
variables n is strictly greater than the number of equations, we must
have $r < n$. Hence there are $n - r$ variables other than $x_{k_1}, \ldots, x_{k_r}$ and
$n - r > 0$. We give these variables arbitrary values, which we can of
course select not all equal to 0. Then we solve for the variables x_{k_r},
$x_{k_{r-1}}, \ldots, x_{k_1}$ starting with the bottom equation and working back up, for
instance

$$x_{k_r} = -S_{k_r}(x)/a_{k_r},$$
$$x_{k_{r-1}} = -S_{k_{r-1}}(x)/a_{k_{r-1}}, \quad \text{and so forth.}$$

This gives us the non-trivial solution, and proves the theorem.

Observe that the pattern follows exactly that of the examples, but with
a notation dealing with the general case.

Exercises II, §4

In each of the following cases find a row equivalent matrix in row echelon form.

1. (a) $\begin{pmatrix} 6 & 3 & -4 \\ -4 & 1 & -6 \\ 1 & 2 & -5 \end{pmatrix}$ (b) $\begin{pmatrix} 1 & 0 & 2 \\ 2 & -1 & 3 \\ 4 & 1 & 8 \end{pmatrix}$.

2. (a) $\begin{pmatrix} 1 & -2 & 3 & -1 \\ 2 & -1 & 2 & 2 \\ 3 & 1 & 2 & 3 \end{pmatrix}$ (b) $\begin{pmatrix} 0 & 1 & 3 & -2 \\ 2 & 1 & -4 & 3 \\ 2 & 3 & 2 & -1 \end{pmatrix}$.

3. (a) $\begin{pmatrix} 1 & 2 & -1 & 2 & 1 \\ 2 & 4 & 1 & -2 & 3 \\ 3 & 6 & 2 & -6 & 5 \end{pmatrix}$ (b) $\begin{pmatrix} 1 & 3 & -1 & 2 \\ 0 & 11 & -5 & 3 \\ 2 & -5 & 3 & 1 \\ 4 & 1 & 1 & 5 \end{pmatrix}$.

4. Write down the coefficient matrix of the linear equations of Exercise 5 in §3, and in each case give a row equivalent matrix in echelon form. Solve the linear equations in each case by this method.

II, §5. Row Operations and Elementary Matrices

Before reading this section, work out the numerical examples given in Exercises 33 through 37 of §2.

The row operations which we used to solve linear equations can be represented by matrix operations. Let $1 \leq r \leq m$ and $1 \leq s \leq m$. Let I_{rs} be the square $m \times m$ matrix which has component 1 in the rs place, and 0 elsewhere:

$$I_{rs} = \begin{pmatrix} 0 & \cdots\cdots\cdots & 0 \\ & \vdots & & \vdots \\ 0 & \cdots 1_{rs} \cdots & 0 \\ & \vdots & & \vdots \\ 0 & \cdots\cdots\cdots & 0 \end{pmatrix}.$$

Let $A = (a_{ij})$ be any $m \times \underline{n}$ matrix. What is the effect of multiplying $I_{rs}A$?

$$r\left\{\begin{pmatrix} 0 & \cdots\cdots\cdots & 0 \\ & \vdots & & \vdots \\ 0 & \cdots 1_{rs} \cdots & 0 \\ & \vdots & & \vdots \\ 0 & \cdots\cdots\cdots & 0 \end{pmatrix}\underbrace{}_{s}\begin{pmatrix} a_{11} & \cdots & a_{1n} \\ & \vdots & \\ & \vdots & \\ a_{s1} & \cdots & a_{sn} \\ & \vdots & \\ a_{m1} & \cdots & a_{mn} \end{pmatrix}\right\}s = \begin{pmatrix} 0 & \cdots & 0 \\ & \vdots & & \vdots \\ a_{s1} & \cdots & a_{sn} \\ & \vdots & & \vdots \\ 0 & \cdots & 0 \end{pmatrix}\right\}r.$$

The definition of multiplication of matrices shows that $I_{rs}A$ is the matrix obtained by putting the s-th row of A in the r-th row, and zeros elsewhere.

If $r = s$ then I_{rr} has a component 1 on the diagonal place, and 0 elsewhere. Multiplication by I_{rr} then leaves the r-th row fixed, and replaces all the other rows by zeros.

If $r \neq s$ let

$$J_{rs} = I_{rs} + I_{sr}.$$

Then

$$J_{rs}A = I_{rs}A + I_{sr}A.$$

Then $I_{rs}A$ puts the s-th row of A in the r-th place, and $I_{sr}A$ puts the r-th row of A in the s-th place. All other rows are replaced by zero. Thus J_{rs} interchanges the r-th row and the s-th row, and replaces all other rows by zero.

Example. Let

$$J = \begin{pmatrix} 0 & 1 & 0 \\ 1 & 0 & 0 \\ 0 & 0 & 0 \end{pmatrix} \quad \text{and} \quad A = \begin{pmatrix} 3 & 2 & -1 \\ 1 & 4 & 2 \\ -2 & 5 & 1 \end{pmatrix}.$$

If you perform the matrix multiplication, you will see directly that JA interchanges the first and second row of A, and replaces the third row by zero.

On the other hand, let

$$E = \begin{pmatrix} 0 & 1 & 0 \\ 1 & 0 & 0 \\ 0 & 0 & 1 \end{pmatrix}$$

Then EA is the matrix obtained from A by interchanging the first and second row, and leaving the third row fixed. We can express E as a sum:

$$E = I_{12} + I_{21} + I_{33}$$

where I_{rs} is the matrix which has rs-component 1, and all other components 0 as before. Observe that E is obtained from the unit matrix by interchanging the first two rows, and leaving the third row unchanged. Thus the operation of interchanging the first two rows of A is carried out by multiplication with the matrix E obtained by doing this operation on the unit matrix.

This is a special case of the following general fact.

Theorem 5.1. *Let E be the matrix obtained from the unit $n \times n$ matrix by interchanging two rows. Let A be an $n \times n$ matrix. Then EA is the matrix obtained from A by interchanging these two rows.*

Proof. The proof is carried out according to the pattern of the example, it is only a question of which symbols are used. Suppose that we interchange the r-th and s-th row. Then we can write

$$E = I_{rs} + I_{sr} + \text{sum of the matrices } I_{jj} \text{ with } j \neq r, j \neq s.$$

Thus E differs from the unit matrix by interchanging the r-th and s-th rows. Then

$$EA = I_{rs}A + I_{sr}A + \text{sum of the matrices } I_{jj}A,$$

with $j \neq r, j \neq s$. By the previous discussion, this is precisely the matrix obtained by interchanging the r-th and s-th rows of A, and leaving all the other rows unchanged.

The same type of discussion also yields the next result.

Theorem 5.2. *Let E be the matrix obtained from the unit $n \times n$ matrix by multiplying the r-th row with a number c and adding it to the s-th row, $r \neq s$. Let A be an $n \times n$ matrix. Then EA is obtained from A by multiplying the r-th row of A by c and adding it to the s-th row of A.*

Proof. We can write

$$E = I + cI_{sr}.$$

Then $EA = A + cI_{sr}A$. We know that $I_{sr}A$ puts the r-th row of A in the s-th place, and multiplication by c multiplies this row by c. All other rows besides the s-th row in $cI_{sr}A$ are equal to 0. Adding $A + cI_{sr}A$ therefore has the effect of adding c times the r-th row of A to the s-th row of A, as was to be shown.

Example. Let

$$E = \begin{pmatrix} 1 & 0 & 4 & 0 \\ 0 & 1 & 0 & 0 \\ 0 & 0 & 1 & 0 \\ 0 & 0 & 0 & 1 \end{pmatrix}.$$

Then E is obtained from the unit matrix by adding 4 times the third row to the first row. Take any $4 \times n$ matrix A and compute EA. You will find that EA is obtained by multiplying the third row of A by 4 and adding it to the first row of A.

More generally, we can let $E_{rs}(c)$ for $r \neq s$ be the elementary matrix.

$$E_{rs}(c) = I + cI_{rs}.$$

$$s \left\{ \begin{pmatrix} 1 & \cdots & 0 & \cdots & 0 & \cdots & 0 \\ \vdots & \ddots & \vdots & & \vdots & & \vdots \\ 0 & \cdots & 1 & \cdots & 0 & \cdots & 0 \\ \vdots & & \vdots & \ddots & \vdots & & \vdots \\ 0 & \cdots & c & \cdots & 1 & \cdots & 0 \\ \vdots & & \vdots & & \vdots & \ddots & \vdots \\ 0 & \cdots & 0 & \cdots & 0 & \cdots & 1 \end{pmatrix} \right. .$$

(with r bracketed over the top)

It differs from the unit matrix by having rs-component equal to c. The effect of multiplication on the left by $E_{rs}(c)$ is to add c times the s-th row to the r-th row.

By an **elementary matrix**, we shall mean any one of the following three types:

(a) A matrix obtained from the unit matrix by multiplying the r-th diagonal component with a number $c \neq 0$.

(b) A matrix obtained from the unit matrix by interchanging two rows (say the r-th and s-th row, $r \neq s$).

(c) A matrix $E_{rs}(c) = I + cI_{rs}$ with $r \neq s$ having rs-component c for $r \neq s$, and all other components 0 except the diagonal components which are equal to 1.

These three types reflect the row operations discussed in the preceding section.

Multiplication by a matrix of type (a) multiplies the r-th row by the number c.

Multiplication by a matrix of type (b) interchanges the r-th and s-th row.

Multiplication by a matrix of type (c) adds c times the s-th row to the r-th row.

Proposition 5.3. *An elementary matrix is invertible.*

Proof. For type (a), the inverse matrix has r-th diagonal component c^{-1}, because multiplying a row first by c and then by c^{-1} leaves the row unchanged.

For type (b), we note that by interchanging the r-th and s-th row twice we return to the same matrix we started with.

For type (c), as in Theorem 5.2, let E be the matrix which adds c times the s-th row to the r-th row of the unit matrix. Let D be the matrix which adds $-c$ times the s-th row to the r-th row of the unit matrix (for $r \neq s$). Then DE is the unit matrix, and so is ED, so E is invertible.

Example. The following elementary matrices are inverse to each other:

$$E = \begin{pmatrix} 1 & 0 & 4 & 0 \\ 0 & 1 & 0 & 0 \\ 0 & 0 & 1 & 0 \\ 0 & 0 & 0 & 1 \end{pmatrix}. \qquad E^{-1} = \begin{pmatrix} 1 & 0 & -4 & 0 \\ 0 & 1 & 0 & 0 \\ 0 & 0 & 1 & 0 \\ 0 & 0 & 0 & 1 \end{pmatrix}.$$

We shall find an effective way of finding the inverse of a square matrix if it has one. This is based on the following properties.

If A, B are square matrices of the same size and have inverses, then so does the product AB, and

$$(AB)^{-1} = B^{-1}A^{-1}.$$

This is immediate, because

$$ABB^{-1}A^{-1} = AIA^{-1} = AA^{-1} = I.$$

Similarly, for any number of factors:

Proposition 5.4. *If $A_1, \dots, A_k$ are invertible matrices of the same size, then their product has an inverse, and*

$$(A_1 \cdots A_k)^{-1} = A_k^{-1} \cdots A_1^{-1}.$$

Note that in the right-hand side, we take the product of the inverses in reverse order. Then

$$A_1 \cdots A_k A_k^{-1} \cdots A_1^{-1} = I$$

because we can collapse $A_k A_k^{-1}$ to I, then $A_{k-1} A_{k-1}^{-1}$ to I and so forth.

Since an elementary matrix has an inverse, we conclude that any product of elementary matrices has an inverse.

Proposition 5.5. *Let A be a square matrix, and let A' be row equivalent to A. Then A has an inverse if and only if A' has an inverse.*

Proof. There exist elementary matrices $E_1, \dots, E_k$ such that

$$A' = E_1 \cdots E_k A.$$

Suppose that A has an inverse. Then the right-hand side has an inverse by Proposition 5.4 since the right-hand side is a product of invertible matrices. Hence A' has an inverse. This proves the proposition.

We are now in a position to find an inverse for a square matrix A if it has one. By Theorem 4.1 we know that A is row equivalent to a matrix A' in echelon form. If one row of A' is zero, then by the definition of echelon form, the last row must be zero, and A' is not invertible, hence A is not invertible. If all the rows of A' are non-zero, then A' is a triangular matrix with non-zero diagonal components. It now suffices to find an inverse for such a matrix. In fact, we prove:

Theorem 5.6. *A square matrix A is invertible if and only if A is row equivalent to the unit matrix. Any upper triangular matrix with non-zero diagonal elements is invertible.*

Proof. Suppose that A is row equivalent to the unit matrix. Then A is invertible by Proposition 5.5. Suppose that A is invertible. We have just seen that A is row equivalent to an upper triangular matrix with non-zero elements on the diagonal. Suppose A is such a matrix:

$$\begin{pmatrix} a_{11} & a_{12} & \cdots & a_{1n} \\ 0 & a_{22} & \cdots & a_{2n} \\ \vdots & \vdots & & \vdots \\ 0 & 0 & \cdots & a_{nn} \end{pmatrix}.$$

By assumption we have $a_{11} \cdots a_{nn} \neq 0$. We multiply the i-th row with a_{ii}^{-1}. We obtain a triangular matrix such that all the diagonal components are equal to 1. Thus to prove the theorem, it suffices to do it in this case, and we may assume that A has the form

$$\begin{pmatrix} 1 & a_{12} & \cdots & a_{1n} \\ 0 & 1 & \cdots & a_{2n} \\ \vdots & \vdots & & \vdots \\ 0 & 0 & \cdots & 1 \end{pmatrix}.$$

We multiply the last row by a_{in} and subtract it from the i-th row for $i = 1, \ldots, n-1$. This makes all the elements of the last column equal to 0 except for the lower right-hand corner, which is 1. We repeat this procedure with the next to the last row, and continue upward. This means that by row equivalences, we can replace all the components which lie strictly above the diagonal by 0. We then terminate with the unit matrix, which is therefore row equivalent with the original matrix. This proves the theorem.

Corollary 5.7. *Let A be an invertible matrix. Then A can be expressed as a product of elementary matrices.*

Proof. This is because A is row equivalent to the unit matrix, and row operations are represented by multiplication with elementary matrices, so there exist $E_1,\ldots,E_k$ such that

$$E_k\cdots E_1 A = I.$$

Then $A = E_1^{-1}\cdots E_k^{-1}$, thus proving the corollary.

When A is so expressed, we also get an expression for the inverse of A, namely

$$A^{-1} = E_k\cdots E_1.$$

The elementary matrices $E_1,\ldots,E_k$ are those which are used to change A to the unit matrix.

Example. Let

$$A = \begin{pmatrix} 2 & -3 & 1 \\ 1 & 1 & -1 \\ 2 & 0 & 1 \end{pmatrix}, \qquad I = \begin{pmatrix} 1 & 0 & 0 \\ 0 & 1 & 0 \\ 0 & 0 & 1 \end{pmatrix}.$$

We want to find an inverse for A. We perform the following row operations, corresponding to the multiplication by elementary matrices as shown.

Interchange first two rows.

$$\begin{pmatrix} 1 & 1 & -1 \\ 2 & -3 & 1 \\ 2 & 0 & 1 \end{pmatrix}, \qquad \begin{pmatrix} 0 & 1 & 0 \\ 1 & 0 & 0 \\ 0 & 0 & 1 \end{pmatrix}.$$

Subtract 2 times first row from second row.
Subtract 2 times first row from third row.

$$\begin{pmatrix} 1 & 1 & -1 \\ 0 & -5 & 3 \\ 0 & -2 & 3 \end{pmatrix}, \qquad \begin{pmatrix} 0 & 1 & 0 \\ 1 & -2 & 0 \\ 0 & -2 & 1 \end{pmatrix}.$$

Subtract 2/5 times second row from third row.

$$\begin{pmatrix} 1 & 1 & -1 \\ 0 & -5 & 3 \\ 0 & 0 & 9/5 \end{pmatrix}, \qquad \begin{pmatrix} 0 & 1 & 0 \\ 1 & -2 & 0 \\ -2/5 & -6/5 & 1 \end{pmatrix}.$$

Subtract 5/3 of third row from second row.
Add 5/9 of third row to first row.

$$\begin{pmatrix} 1 & 1 & 0 \\ 0 & -5 & 0 \\ 0 & 0 & 9/5 \end{pmatrix}, \qquad \begin{pmatrix} -2/9 & 1/3 & 5/9 \\ 5/3 & 0 & -5/3 \\ -2/5 & -6/5 & 1 \end{pmatrix}.$$

Add 1/5 of second row to first row.

$$\begin{pmatrix} 1 & 0 & 0 \\ 0 & -5 & 0 \\ 0 & 0 & 9/5 \end{pmatrix}, \qquad \begin{pmatrix} 1/9 & 1/3 & 2/9 \\ 5/3 & 0 & -5/3 \\ -2/5 & -6/5 & 1 \end{pmatrix}.$$

Multiply second row by $-1/5$.
Multiply third row by 5/9.

$$\begin{pmatrix} 1 & 0 & 0 \\ 0 & 1 & 0 \\ 0 & 0 & 1 \end{pmatrix}, \qquad \begin{pmatrix} 1/9 & 1/3 & 2/9 \\ -1/3 & 0 & 1/3 \\ -2/9 & -2/3 & 5/9 \end{pmatrix}.$$

Then A^{-1} is the matrix on the right, that is

$$A^{-1} = \begin{pmatrix} 1/9 & 1/3 & 2/9 \\ -1/3 & 0 & 1/3 \\ -2/9 & -2/3 & 5/9 \end{pmatrix}.$$

You can check this by direct multiplication with A to find the unit matrix.

If A is a square matrix and we consider an inhomogeneous system of linear equations

$$AX = B,$$

then we can use the inverse to solve the system, if A is invertible. Indeed, in this case, we multiply both sides on the left by A^{-1} and we find

$$X = A^{-1}B.$$

This also proves:

Proposition 5.8. *Let $AX = B$ be a system of n linear equations in n unknowns. Assume that the matrix of coefficients A is invertible. Then there is a unique solution X to the system, and*

$$X = A^{-1}B.$$

Exercises II, §5

1. Using elementary row operations, find inverses for the following matrices.

(a) $\begin{pmatrix} 2 & 1 & 2 \\ 0 & 3 & -1 \\ 4 & 1 & 1 \end{pmatrix}$ (b) $\begin{pmatrix} 3 & -1 & 5 \\ -1 & 2 & 1 \\ -2 & 4 & 3 \end{pmatrix}$

(c) $\begin{pmatrix} 2 & 4 & 3 \\ -1 & 3 & 0 \\ 0 & 2 & 1 \end{pmatrix}$ (d) $\begin{pmatrix} 1 & 2 & -1 \\ 0 & 1 & 1 \\ 0 & 2 & 7 \end{pmatrix}$

(e) $\begin{pmatrix} -1 & 5 & 3 \\ 4 & 0 & 0 \\ 2 & 7 & 8 \end{pmatrix}$ (f) $\begin{pmatrix} 3 & 1 & 2 \\ 4 & 5 & 1 \\ -1 & 2 & -1 \end{pmatrix}$

Note: For another way of finding inverses, see the chapter on determinants.

2. Let $r \neq s$. Show that $I_{rs}^2 = O$.

3. Let $r \neq s$. Let $E_{rs}(c) = I + cI_{rs}$. Show that

$$E_{rs}(c)E_{rs}(c') = E_{rs}(c + c').$$

II, §6. Linear Combinations

Let $A^1, \ldots, A^n$ be m-tuples in $\mathbf{R}^m$. Let $x_1, \ldots, x_n$ be numbers. Then we call

$$x_1 A^1 + \cdots + x_n A^n$$

a **linear combination** of $A^1, \ldots, A^n$; and we call $x_1, \ldots, x_n$ the **coefficients** of the linear combination. A similar definition applies to a linear combination of row vectors.

The linear combination is called **non-trivial** if not all the coefficients $x_1, \ldots, x_n$ are equal to 0.

Consider once more a system of linear homogeneous equations

$$(**) \qquad \begin{matrix} a_{11}x_1 + \cdots + a_{1n}x_n = 0 \\ \vdots \qquad\qquad \vdots \\ a_{m1}x_1 + \cdots + a_{mn}x_n = 0. \end{matrix}$$

Our system of homogeneous equations can also be written in the form

$$x_1 \begin{pmatrix} a_{11} \\ a_{21} \\ \vdots \\ a_{m1} \end{pmatrix} + x_2 \begin{pmatrix} a_{12} \\ a_{22} \\ \vdots \\ a_{m2} \end{pmatrix} + \cdots + x_n \begin{pmatrix} a_{1n} \\ a_{2n} \\ \vdots \\ a_{mn} \end{pmatrix} = \begin{pmatrix} 0 \\ 0 \\ \vdots \\ 0 \end{pmatrix},$$

or more concisely:

$$x_1 A^1 + \cdots + x_n A^n = O,$$

where $A^1, \ldots, A^n$ are the column vectors of the matrix of coefficients, which is $A = (a_{ij})$. Thus the problem of finding a non-trivial solution for the system of homogeneous linear equations is equivalent to finding a non-trivial linear combination of $A^1, \ldots, A^n$ which is equal to O.

Vectors $A^1, \ldots, A^n$ are called **linearly dependent** if there exist numbers $x_1, \ldots, x_n$ not all equal to 0 such that

$$x_1 A^1 + \cdots + x_n A^n = O.$$

Thus a non-trivial solution $(x_1, \ldots, x_n)$ is an n-tuple which gives a linear combination of $A^1, \ldots, A^n$ equal to O, i.e. a relation of linear dependence between the columns of A. *We may thus summarize the description of the set of solutions of the system of homogeneous linear equations in a table.*

(a) *It consists of those vectors X giving linear relations*

$$x_1 A^1 + \cdots + x_n A^n = O$$

between the columns of A.

(b) *It consists of those vectors X perpendicular to the rows of A, that is $X \cdot A_i = 0$ for all i.*

(c) *It consists of those vectors X such that $AX = O$.*

Vectors $A^1, \ldots, A^n$ are called **linearly independent** if, given any linear combination of them which is equal to O, i.e.

$$x_1 A^1 + \cdots + x_n A^n = O,$$

then we must necessarily have $x_j = 0$ for all $j = 1, \ldots, n$. This means that there is no non-trivial relation of linear dependence among the vectors $A^1, \ldots, A^n$.

Example. The standard unit vectors

$$E_1 = (1, 0, \ldots, 0), \ldots, E_n = (0, \ldots, 0, 1)$$

of $\mathbf{R}^n$ are linearly independent. Indeed, let $x_1, \ldots, x_n$ be numbers such that

$$x_1 E_1 + \cdots + x_n E_n = O.$$

The left-hand side is just the n-tuple $(x_1, \ldots, x_n)$. If this n-tuple is O, then all components are 0, so $x_i = 0$ for all i. This proves that $E_1, \ldots, E_n$ are linearly independent.

We shall study the notions of linear dependence and independence more systematically in the next chapter. They were mentioned here just to have a complete table for the three basic interpretations of a system of linear equations, and to introduce the notion in a concrete special case before giving the general definitions in vector spaces.

Exercise II, §6

1. (a) Let $A = (a_{ij})$, $B = (b_{jk})$ and let $AB = C$ with $C = (c_{ik})$. Let C^k be the k-th column of C. Express C^k as a linear combination of the columns of A. Describe precisely which are the coefficients, coming from the matrix B.
   (b) Let $AX = C^k$ where X is some column of B. Which column is it?

CHAPTER III

Vector Spaces

As usual, a collection of objects will be called a **set**. A member of the collection is also called an **element** of the set. It is useful in practice to use short symbols to denote certain sets. For instance we denote by **R** the set of all numbers. To say that "x is a number" or that "x is an element of **R**" amounts to the same thing. The set of n-tuples of numbers will be denoted by $\mathbf{R}^n$. Thus "X is an element of $\mathbf{R}^n$" and "X is an n-tuple" mean the same thing. Instead of saying that u is an element of a set S, we shall also frequently say that u **lies in** S and we write $u \in S$. If S and S' are two sets, and if every element of S' is an element of S, then we say that S' is a **subset** of S. Thus the set of rational numbers is a subset of the set of (real) numbers. To say that S is a subset of S' is to say that S is part of S'. To denote the fact that S is a subset of S', we write $S \subset S'$.

If S_1, S_2 are sets, then the **intersection** of S_1 and S_2, denoted by $S_1 \cap S_2$, is the set of elements which lie in both S_1 and S_2. The **union** of S_1 and S_2, denoted by $S_1 \cup S_2$, is the set of elements which lie in S_1 or S_2.

III, §1. Definitions

In mathematics, we meet several types of objects which can be added and multiplied by numbers. Among these are vectors (of the same dimension) and functions. It is now convenient to define in general a notion which includes these as a special case.

A **vector space** V is a set of objects which can be added and multiplied by numbers, in such a way that the sum of two elements of V is again an element of V, the product of an element of V by a number is an element of V, and the following properties are satisfied:

VS 1. *Given the elements* u, v, w *of* V, *we have*

$$(u + v) + w = u + (v + w).$$

VS 2. *There is an element of* V, *denoted by* O, *such that*

$$O + u = u + O = u$$

for all elements u *of* V.

VS 3. *Given an element* u *of* V, *the element* $(-1)u$ *is such that*

$$u + (-1)u = O.$$

VS 4. *For all elements* u, v *of* V, *we have*

$$u + v = v + u.$$

VS 5. *If* c *is a number, then* $c(u + v) = cu + cv$.

VS 6. *If* a, b *are two numbers, then* $(a + b)v = av + bv$.

VS 7. *If* a, b *are two numbers, then* $(ab)v = a(bv)$.

VS 8. *For all elements* u *of* V, *we have* $1 \cdot u = u$ (1 *here is the number one*).

We have used all these rules when dealing with vectors, or with functions but we wish to be more systematic from now on, and hence have made a list of them. Further properties which can be easily deduced from these are given in the exercises and will be assumed from now on.

The algebraic properties of elements of an arbitrary vector space are very similar to those of elements of $\mathbf{R}^2$, $\mathbf{R}^3$, or $\mathbf{R}^n$. Consequently it is customary to call elements of an arbitrary vector space also **vectors**.

If u, v are vectors (i.e. elements of the arbitrary vector space V), then the sum

$$u + (-1)v$$

is usually written $u - v$. We also write $-v$ instead of $(-1)v$.

Example 1. Fix two positive integers m, n. Let V be the set of all $m \times n$ matrices. We also denote V by Mat($m \times n$). Then V is a vector space. It is easy to verify that all properties **VS 1** through **VS 8** are satisfied by our rules for addition of matrices and multiplication of matrices by numbers. The main thing to observe here is that addition of matrices is defined in terms of the components, and for the addition of components, the conditions analogous to **VS 1** through **VS 4** are satisfied. They are standard properties of numbers. Similarly, **VS 5** through **VS 8** are true for multiplication of matrices by numbers, because the corresponding properties for the multiplication of numbers are true.

Example 2. Let V be the set of all functions defined for all numbers. If f, g are two functions, then we know how to form their sum $f + g$. It is the function whose value at a number t is $f(t) + g(t)$. We also know how to multiply f by a number c. It is the function cf whose values at a number t is $cf(t)$. In dealing with functions, we have used properties **VS 1** through **VS 8** many times. We now realize that the set of functions is a vector space.

The function f such that $f(t) = 0$ **for all** t is the **zero function**. We emphasize the condition for all t. If a function has some of its values equal to zero, but other values not equal to 0, then it is **not** the zero function.

In practice, a number of elementary properties concerning addition of elements in a vector space are obvious because of the concrete way the vector space is given in terms of numbers, for instance as in the previous two examples. We shall now see briefly how to prove such properties just from the axioms.

It is possible to add several elements of a vector space. Suppose we wish to add four elements, say u, v, w, z. We first add any two of them, then a third, and finally a fourth. Using the rules **VS 1** and **VS 4**, we see that it does not matter in which order we perform the additions. This is exactly the same situation as we had with vectors. For example, we have

$$((u + v) + w) + z = (u + (v + w)) + z$$
$$= ((v + w) + u) + z$$
$$= (v + w) + (u + z), \quad \text{etc.}$$

Thus it is customary to leave out the parentheses, and write simply

$$u + v + w + z.$$

The same remark applies to the sum of any number n of elements of V.

We shall use 0 to denote the number zero, and O to denote the element of any vector space V satisfying property **VS 2**. We also call it zero, but there is never any possibility of confusion. We observe that this zero element O is uniquely determined by condition **VS 2**. Indeed, if

$$v + w = v$$

then adding $-v$ to both sides yields

$$-v + v + w = -v + v = O,$$

and the left-hand side is just $O + w = w$, so $w = O$.

Observe that for any element v in V we have

$$Ov = O.$$

Proof.

$$O = v + (-1)v = (1 - 1)v = Ov.$$

Similarly, if c is a number, then

$$cO = O.$$

Proof. We have $cO = c(O + O) = cO + cO$. Add $-cO$ to both sides to get $cO = O$.

Subspaces

Let V be a vector space, and let W be a subset of V. Assume that W satisfies the following conditions.

(i)   If v, w are elements of W, their sum $v + w$ is also an element of W.

(ii) If v is an element of W and c a number, then cv is an element of W.

(iii) The element O of V is also an element of W.

Then W itself is a vector space. Indeed, properties **VS 1** through **VS 8**, being satisfied for all elements of V, are satisfied also for the elements of W. We shall call W a **subspace** of V.

Example 3. Let $V = \mathbf{R}^n$ and let W be the set of vectors in V whose last coordinate is equal to 0. Then W is a subspace of V, which we could identify with $\mathbf{R}^{n-1}$.

Example 4. Let A be a vector in $\mathbf{R}^3$. Let W be the set of all elements B in $\mathbf{R}^3$ such that $B \cdot A = 0$, i.e. such that B is perpendicular to A. Then W is a subspace of $\mathbf{R}^3$. To see this, note that $O \cdot A = 0$, so that O is in W. Next, suppose that B, C are perpendicular to A. Then

$$(B + C) \cdot A = B \cdot A + C \cdot A = 0,$$

so that $B + C$ is also perpendicular to A. Finally, if x is a number, then

$$(xB) \cdot A = x(B \cdot A) = 0,$$

so that xB is perpendicular to A. This proves that W is a subspace of $\mathbf{R}^3$.

More generally, if A is a vector in $\mathbf{R}^n$, then the set of all elements B in $\mathbf{R}^n$ such that $B \cdot A = 0$ is a subspace of $\mathbf{R}^n$. The proof is the same as when $n = 3$.

Example 5. Let Sym($n \times n$) be the set of all symmetric $n \times n$ matrices. Then Sym($n \times n$) is a subspace of the space of all $n \times n$ matrices. Indeed, if A, B are symmetric and c is a number, then $A + B$ and cA are symmetric. Also the zero matrix is symmetric.

Example 6. If f, g are two continuous functions, then $f + g$ is continuous. If c is a number, then cf is continuous. The zero function is continuous. Hence the continuous functions form a subspace of the vector space of all functions.

If f, g are two differentiable functions, then their sum $f + g$ is differentiable. If c is a number, then cf is differentiable. The zero function is differentiable. Hence the differentiable functions form a subspace of the vector space of all functions. Furthermore, every differentiable function is continuous. Hence the differentiable functions form a subspace of the vector space of continuous functions.

Example 7. Let V be a vector space and let U, W be subspaces. We denote by $U \cap W$ the **intersection** of U and W, i.e. the set of elements which lie both in U and W. Then $U \cap W$ is a subspace. For instance, if U, W are two planes in 3-space passing through the origin, then in general, their intersection will be a straight line passing through the origin, as shown in Fig. 1.

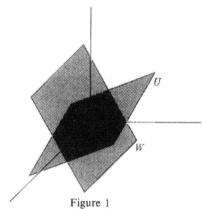

Figure 1

Example 8. Let U, W be subspaces of a vector space V. By

$$U + W$$

we denote the set of all elements $u + w$ with $u \in U$ and $w \in W$. Then we leave it to the reader to verify that $U + W$ is a subspace of V, said to be **generated by** U and W, and called the **sum** of U and W.

Exercises III, §1

1. Let $A_1, \ldots, A_r$ be vectors in $\mathbf{R}^n$. Let W be the set of vectors B in $\mathbf{R}^n$ such that $B \cdot A_i = 0$ for every $i = 1, \ldots, r$. Show that W is a subspace of $\mathbf{R}^n$.

2. Show that the following sets of elements in $\mathbf{R}^2$ form subspaces.
   (a) The set of all (x, y) such that $x = y$.
   (b) The set of all (x, y) such that $x - y = 0$.
   (c) The set of all (x, y) such that $x + 4y = 0$.

3. Show that the following sets of elements in $\mathbf{R}^3$ form subspaces.
   (a) The set of all (x, y, z) such that $x + y + z = 0$.
   (b) The set of all (x, y, z) such that $x = y$ and $2y = z$.
   (c) The set of all (x, y, z) such that $x + y = 3z$.

4. If U, W are subspaces of a vector space V, show that $U \cap W$ and $U + W$ are subspaces.

5. Let V be a subspace of $\mathbf{R}^n$. Let W be the set of elements of $\mathbf{R}^n$ which are perpendicular to every element of V. Show that W is a subspace of $\mathbf{R}^n$. This subspace W is often denoted by $V^{\perp}$, and is called V **perp**, or also the **orthogonal complement** of V.

III, §2. Linear Combinations

Let V be a vector space, and let $v_1, \ldots, v_n$ be elements of V. We shall say that $v_1, \ldots, v_n$ **generate** V if given an element $v \in V$ there exist numbers $x_1, \ldots, x_n$ such that

$$v = x_1 v_1 + \cdots + x_n v_n.$$

Example 1. Let $E_1, \ldots, E_n$ be the standard unit vectors in $\mathbf{R}^n$, so E_i has component 1 in the i-th place, and component 0 in all other places.

Then $E_1, \ldots, E_n$ generate $\mathbf{R}^n$. Proof: given $X = (x_1, \ldots, x_n) \in \mathbf{R}^n$. Then

$$X = \sum_{i=1}^{n} x_i E_i,$$

so there exist numbers satisfying the condition of the definition.

Let V be an arbitrary vector space, and let $v_1, \ldots, v_n$ be elements of V. Let $x_1, \ldots, x_n$ be numbers. An expression of type

$$x_1 v_1 + \cdots + x_n v_n$$

is called a **linear combination** of $v_1, \ldots, v_n$. The numbers $x_1, \ldots, x_n$ are then called the **coefficients** of the linear combination.

The set of all linear combinations of $v_1, \ldots, v_n$ is a subspace of V.

Proof. Let W be the set of all such linear combinations. Let $y_1, \ldots, y_n$ be numbers. Then

$$(x_1 v_1 + \cdots + x_n v_n) + (y_1 v_1 + \cdots + y_n v_n)$$
$$= (x_1 + y_1) v_1 + \cdots + (x_n + y_n) v_n.$$

Thus the sum of two elements of W is again an element of W, i.e. a linear combination of $v_1, \ldots, v_n$. Furthermore, if c is a number, then

$$c(x_1 v_1 + \cdots + x_n v_n) = c x_1 v_1 + \cdots + c x_n v_n$$

is a linear combination of $v_1, \ldots, v_n$, and hence is an element of W. Finally,

$$O = 0 v_1 + \cdots + 0 v_n$$

is an element of W. This proves that W is a subspace of V.

The subspace W consisting of all linear combinations of $v_1, \ldots, v_n$ is called the subspace **generated** by $v_1, \ldots, v_n$.

Example 2. Let v_1 be a non-zero element of a vector space V, and let w be any element of V. The set of elements

$$w + t v_1 \qquad \text{with} \qquad t \in \mathbf{R}$$

is called the **line passing through** w **in the direction of** v_1. We have already met such lines in Chapter I, §5. If $w = O$, then the line consisting of all scalar multiples tv_1 with $t \in \mathbf{R}$ is a subspace, generated by v_1.

Let v_1, v_2 be elements of a vector space V, and assume that neither is a scalar multiple of the other. The subspace generated by v_1, v_2 is called the **plane** generated by v_1, v_2. It consists of all linear combinations

$$t_1 v_1 + t_2 v_2 \qquad \text{with} \qquad t_1, t_2 \text{ arbitrary numbers.}$$

This plane passes through the origin, as one sees by putting $t_1 = t_2 = 0$.

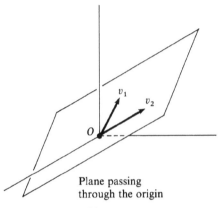

Plane passing
through the origin

Figure 2

We obtain the most general notion of a plane by the following operation. Let S be an arbitrary subset of V. Let P be an element of V. If we add P to all elements of S, then we obtain what is called the **translation** of S by P. It consists of all elements $P + v$ with v in S.

Example 3. Let v_1, v_2 be elements of a vector space V such that neither is a scalar multiple of the other. Let P be an element of V. We define the **plane passing through** P, **parallel to** v_1, v_2 to be the set of all elements

$$P + t_1 v_1 + t_2 v_2$$

where t_1, t_2 are arbitrary numbers. This notion of plane is the analogue, with two elements v_1, v_2, of the notion of parametrized line considered in Chapter I.

Warning. Usually such a plane does not pass through the origin, as shown on Fig. 3. Thus such a plane is **not** a subspace of V. If we take $P = O$, however, then the plane is a subspace.

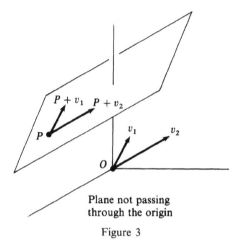

Plane not passing
through the origin

Figure 3

Sometimes it is interesting to restrict the coefficients of a linear combination. We give a number of examples below.

Example 4. Let V be a vector space and let v, u be elements of V. We define the **line segment** between v and $v + u$ to be the set of all points

$$v + tu, \qquad 0 \leqq t \leqq 1.$$

This line segment is illustrated in the following picture.

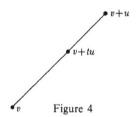

Figure 4

For instance, if $t = \frac{1}{2}$, then $v + \frac{1}{2}u$ is the point midway between v and $v + u$. Similarly, if $t = \frac{1}{3}$, then $v + \frac{1}{3}u$ is the point one third of the way between v and $v + u$ (Fig. 5).

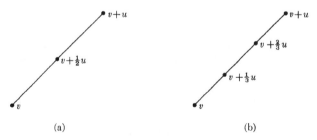

(a)   (b)

Figure 5

If v, w are elements of V, let $u = w - v$. Then the **line segment between** v and w is the set of all points $v + tu$, or

$$v + t(w - v), \qquad 0 \leq t \leq 1.$$

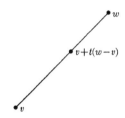

Figure 6

Observe that we can rewrite the expression for these points in the form

(1) $$(1 - t)v + tw, \qquad 0 \leq t \leq 1,$$

and letting $s = 1 - t$, $t = 1 - s$, we can also write it as

$$sv + (1 - s)w, \qquad 0 \leq s \leq 1.$$

Finally, we can write the points of our line segment in the form

(2) $$t_1 v + t_2 w \qquad \text{with} \quad t_1, t_2 \geq 0 \quad \text{and} \quad t_1 + t_2 = 1.$$

Indeed, letting $t = t_2$, we see that every point which can be written in the form (2) satisfies (1). Conversely, we let $t_1 = 1 - t$ and $t_2 = t$ and see that every point of the form (1) can be written in the form (2).

Example 5. Let v, w be elements of a vector space V. Assume that neither is a scalar multiple of the other. We define the **parallelogram spanned** by v, w to be the set of all points

$$t_1 v + t_2 w, \qquad 0 \leq t_i \leq 1 \quad \text{for} \quad i = 1, 2.$$

This definition is clearly justified since $t_1 v$ is a point of the segment between O and v (Fig. 7), and $t_2 w$ is a point of the segment between O

and w. For all values of t_1, t_2 ranging independently between 0 and 1, we see geometrically that $t_1 v + t_2 w$ describes all points of the parallelogram.

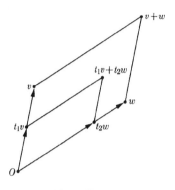

Figure 7

We obtain the most general parallelogram (Fig. 8) by taking the translation of the parallelogram just described. Thus if u is an element of V, the translation by u of the parallelogram spanned by v and w consists of all points

$$u + t_1 v + t_2 w, \qquad 0 \leqq t_i \leqq 1 \quad \text{for} \quad i = 1, 2.$$

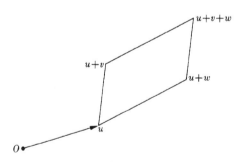

Figure 8

Similarly, in higher dimensions, let v_1, v_2, v_3 be elements of a vector space V. We define the **box spanned by these elements** to be the set of linear combinations

$$t_1 v_1 + t_2 v_2 + t_3 v_3 \qquad \text{with} \qquad 0 \leqq t_i \leqq 1.$$

We draw the picture when v_1, v_2, v_3 are in general position:

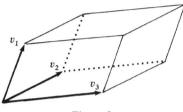

Figure 9

There may be degenerate cases, which will lead us into the notion of linear dependence a little later.

Exercises III, §2

1. Let $A_1,\dots,A_r$ be generators of a subspace V of $\mathbf{R}^n$. Let W be the set of all elements of $\mathbf{R}^n$ which are perpendicular to $A_1,\dots,A_r$. Show that the vectors of W are perpendicular to every element of V.

2. Draw the parallelogram spanned by the vectors $(1, 2)$ and $(-1, 1)$ in $\mathbf{R}^2$.

3. Draw the parallelogram spanned by the vectors $(2, -1)$ and $(1, 3)$ in $\mathbf{R}^2$.

III, §3. Convex Sets

Let S be a subset of a vector space V. We shall say that S is **convex** if given points P, Q in S then the line segment between P and Q is contained in S. In Fig. 10, the set on the left is convex. The set on the right is not convex since the line segment between P and Q is not entirely contained in S.

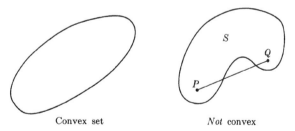

Convex set   *Not* convex

Figure 10

We recall that the line segment between P and Q consists of all points

$$(1 - t)P + tQ \qquad \text{with} \qquad 0 \leq t \leq 1.$$

This gives us a simple test to determine whether a set is convex or not.

Example 1. Let S be the parallelogram spanned by two vectors v_1, v_2, so S is the set of linear combinations

$$t_1 v_1 + t_2 v_2 \qquad \text{with} \qquad 0 \leq t_i \leq 1.$$

We wish to prove that S is convex. Let

$$P = t_1 v_1 + t_2 v_2 \qquad \text{and} \qquad Q = s_1 v_1 + s_2 v_2$$

be points in S. Then

$$(1 - t)P + tQ = (1 - t)(t_1 v_1 + t_2 v_2) + t(s_1 v_1 + s_2 v_2)$$
$$= (1 - t)t_1 v_1 + (1 - t)t_2 v_2 + ts_1 v_1 + ts_2 v_2$$
$$= r_1 v_1 + r_2 v_2,$$

where

$$r_1 = (1 - t)t_1 + ts_1 \qquad \text{and} \qquad r_2 = (1 - t)t_2 + ts_2.$$

But we have

$$0 \leq (1 - t)t_1 + ts_1 \leq (1 - t) + t = 1$$

and

$$0 \leq (1 - t)t_2 + ts_2 \leq (1 - t) + t = 1.$$

Hence

$$(1 - t)P + tQ = r_1 v_1 + r_2 v_2 \qquad \text{with} \qquad 0 \leq r_i \leq 1.$$

This proves that $(1 - t)P + tQ$ is in the parallelogram, which is therefore convex.

Example 2. Half planes. Consider a linear equation like

$$2x - 3y = 6.$$

This is the equation of a line as shown on Fig. 11.

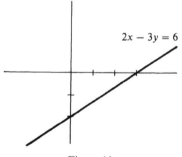

$$2x - 3y = 6$$

Figure 11

The inequalities

$$2x - 3y \leqq 6 \qquad \text{and} \qquad 2x - 3y \geqq 6$$

determine two half planes; one of them lies below the line and the other lies above the line, as shown on Fig. 12.

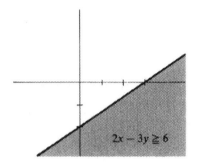
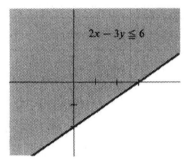

Figure 12

Let $A = (2, -3)$. We can, and should write the linear inequalities in the form

$$A \cdot X \geqq 6 \qquad \text{and} \qquad A \cdot X \leqq 6,$$

where $X = (x, y)$. Prove as Exercise 2 that each half plane is convex. This is clear intuitively from the picture, at least in $\mathbf{R}^2$, but your proof should be valid for the analogous situation in $\mathbf{R}^n$.

Theorem 3.1. *Let $P_1, \ldots, P_n$ be points of a vector space V. Let S be the set of all linear combinations*

$$t_1 P_1 + \cdots + t_n P_n$$

with $0 \leqq t_i$ and $t_1 + \cdots + t_n = 1$. Then S is convex.

Proof. Let

$$P = t_1 P_1 + \cdots + t_n P_n$$

and

$$Q = s_1 P_1 + \cdots + s_n P_n$$

with $0 \leqq t_i$, $0 \leqq s_i$, and

$$t_1 + \cdots + t_n = 1,$$
$$s_1 + \cdots + s_n = 1.$$

Let $0 \leq t \leq 1$. Then:

$$(1 - t)P + tQ = (1 - t)t_1 P_1 + \cdots + (1 - t)t_n P_n$$
$$+ ts_1 P_1 + \cdots + ts_n P_n$$
$$= [(1 - t)t_1 + ts_1]P_1 + \cdots + [(1 - t)t_n + ts_n]P_n.$$

We have $0 \leq (1 - t)t_i + ts_i$ for all i, and

$$(1 - t)t_1 + ts_1 + \cdots + (1 - t)t_n + ts_n$$
$$= (1 - t)(t_1 + \cdots + t_n) + t(s_1 + \cdots + s_n)$$
$$= (1 - t) + t$$
$$= 1.$$

This proves our theorem.

In the next theorem, we shall prove that the set of all linear combinations

$$t_1 P_1 + \cdots + t_n P_n \qquad \text{with} \qquad 0 \leq t_i \quad \text{and} \quad t_1 + \cdots + t_n = 1$$

is the smallest convex set containing $P_1, \ldots, P_n$. For example, suppose that P_1, P_2, P_3 are three points in the plane not on a line. Then it is geometrically clear that the smallest convex set containing these three points is the triangle having these points as vertices.

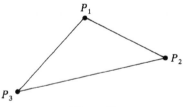

Figure 13

Thus it is natural to take as definition of a triangle the following property, valid in any vector space.

Let P_1, P_2, P_3 be three points in a vector space V, not lying on a line. Then the **triangle spanned** by these points is the set of all combinations

$$t_1 P_1 + t_2 P_2 + t_3 P_3 \qquad \text{with} \qquad 0 \leq t_i \quad \text{and} \quad t_1 + t_2 + t_3 = 1.$$

When we deal with more than three points, then the set of linear combinations as in Theorem 3.1 looks as in the following figure.

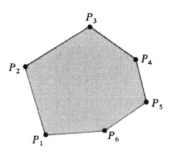

Figure 14

We shall call the convex set of Theorem 3.1 the convex set **spanned** by $P_1, \ldots, P_n$. Although we shall not need the next result, it shows that this convex set is the smallest convex set containing all the points $P_1, \ldots, P_n$. Omit the proof if you can't handle the argument by induction.

Theorem 3.2. *Let $P_1, \ldots, P_n$ be points of a vector space V. Any convex set which contains $P_1, \ldots, P_n$ also contains all linear combinations*

$$t_1 P_1 + \cdots + t_n P_n$$

with $0 \leq t_i$ for all i and $t_1 + \cdots + t_n = 1$.

Proof. We prove this by induction. If $n = 1$, then $t_1 = 1$, and our assertion is obvious. Assume the theorem proved for some integer $n - 1 \geq 1$. We shall prove it for n. Let $t_1, \ldots, t_n$ be numbers satisfying the conditions of the theorem. Let S' be a convex set containing $P_1, \ldots, P_n$. We must show that S' contains all linear combinations

$$t_1 P_1 + \cdots + t_n P_n.$$

If $t_n = 1$, then our assertion is trivial because $t_1 = \cdots = t_{n-1} = 0$. Suppose that $t_n \neq 1$. Then the linear combination $t_1 P_1 + \cdots + t_n P_n$ is equal to

$$(1 - t_n)\left(\frac{t_1}{1 - t_n} P_1 + \cdots + \frac{t_{n-1}}{1 - t_n} P_{n-1} \right) + t_n P_n.$$

Let

$$s_i = \frac{t_i}{1 - t_i} \qquad \text{for} \qquad i = 1, \ldots, n - 1.$$

Then $s_i \geq 0$ and $s_1 + \cdots + s_{n-1} = 1$ so that by induction, we conclude that the point

$$Q = s_1 P_1 + \cdots + s_{n-1} P_{n-1}$$

lies in S'. But then

$$(1 - t_n)Q + t_n P_n = t_1 P_1 + \cdots + t_n P_n$$

lies in S' by definition of a convex set, as was to be shown.

Exercises III, §3

1. Let S be the parallelogram consisting of all linear combinations $t_1 v_1 + t_2 v_2$ with $0 \le t_1 \le 1$ and $0 \le t_2 \le 1$. Prove that S is convex.

2. Let A be a non-zero vector in $\mathbf{R}^n$ and let c be a fixed number. Show that the set of all elements X in $\mathbf{R}^n$ such that $A \cdot X \ge c$ is convex.

3. Let S be a convex set in a vector space. If c is a number, denote by cS the set of all elements cv with v in S. Show that cS is convex.

4. Let S_1 and S_2 be convex sets. Show that the intersection $S_1 \cap S_2$ is convex.

5. Let S be a convex set in a vector space V. Let w be an arbitrary element of V. Let $w + S$ be the set of all elements $w + v$ with v in S. Show that $w + S$ is convex.

III, §4. Linear Independence

Let V be a vector space, and let $v_1, \ldots, v_n$ be elements of V. We shall say that $v_1, \ldots, v_n$ are **linearly dependent** if there exist numbers $a_1, \ldots, a_n$ not all equal to 0 such that

$$a_1 v_1 + \cdots + a_n v_n = O.$$

If there do not exist such numbers, then we say that $v_1, \ldots, v_n$ are **linearly independent.** In other words, vectors $v_1, \ldots, v_n$ are linearly independent if and only if the following condition is satisfied:

Let $a_1, \ldots, a_n$ be numbers such that

$$a_1 v_1 + \cdots + a_n v_n = O;$$

then $a_i = 0$ for all $i = 1, \ldots, n$.

Example 1. Let $V = \mathbf{R}^n$ and consider the vectors

$$E_1 = (1, 0, \ldots, 0)$$
$$\vdots$$
$$E_n = (0, 0, \ldots, 1).$$

Then $E_1,\dots,E_n$ are linearly independent. Indeed, let $a_1,\dots,a_n$ be numbers such that $a_1E_1 + \cdots + a_nE_n = O$. Since

$$a_1E_1 + \cdots + a_nE_n = (a_1,\dots,a_n),$$

it follows that all $a_i = 0$.

Example 2. Show that the vectors $(1, 1)$ and $(-3, 2)$ are linearly independent.

Let a, b be two numbers such that

$$a(1, 1) + b(-3, 2) = O.$$

Writing this equation in terms of components, we find

$$a - 3b = 0, \qquad a + 2b = 0.$$

This is a system of two equations which we solve for a and b. Subtracting the second from the first, we get $-5b = 0$, whence $b = 0$. Substituting in either equation, we find $a = 0$. Hence, a, b are both 0, and our vectors are linearly independent.

If elements $v_1,\dots,v_n$ of V generate V and in addition are linearly independent, then $\{v_1,\dots,v_n\}$ is called a **basis** of V. We shall also say that the elements $v_1,\dots,v_n$ **constitute** or **form** a basis of V.

Example 3. The vectors $E_1,\dots,E_n$ of Example 1 form a basis of $\mathbf{R}^n$. To prove this we have to prove that they are linearly independent, which was already done in Example 1; and that they generate $\mathbf{R}^n$. Given an element $A = (a_1,\dots,a_n)$ of $\mathbf{R}^n$ we can write A as a linear combination

$$A = a_1E_1 + \cdots + a_nE_n,$$

so by definition, $E_1,\dots,E_n$ generate $\mathbf{R}^n$. Hence they form a basis.

However, there are many other bases. Let us look at $n = 2$. We shall find out that any two vectors which are not parallel form a basis of $\mathbf{R}^2$. Let us first consider an example.

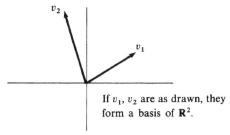

If v_1, v_2 are as drawn, they
form a basis of $\mathbf{R}^2$.

Figure 15

Example 4. Show that the vectors $(1, 1)$ and $(-1, 2)$ form a basis of $\mathbf{R}^2$.

We have to show that they are linearly independent and that they generate $\mathbf{R}^2$. To prove linear independence, suppose that a, b are numbers such that

$$a(1, 1) + b(-1, 2) = (0, 0)$$

Then

$$a - b = 0, \qquad a + 2b = 0.$$

Subtracting the first equation from the second yields $3b = 0$, so that $b = 0$. But then from the first equation, $a = 0$, thus proving that our vectors are linearly independent.

Next, we must show that $(1, 1)$ and $(-1, 2)$ generate $\mathbf{R}^2$. Let (s, t) be an arbitrary element of $\mathbf{R}^2$. We have to show that there exist numbers x, y such that

$$x(1, 1) + y(-1, 2) = (s, t).$$

In other words, we must solve the system of equations

$$x - y = s,$$
$$x + 2y = t.$$

Again subtract the first equation from the second. We find

$$3y = t - s,$$

whence

$$y = \frac{t - s}{3},$$

and finally

$$x = y + s = \frac{t - s}{3} + s.$$

This proves that $(1, 1)$ and $(-1, 2)$ generate $\mathbf{R}^2$, and concludes the proof that they form a basis of $\mathbf{R}^2$.

The general story for $\mathbf{R}^2$ is expressed in the following theorem.

Theorem 4.1. *Let (a, b) and (c, d) be two vectors in $\mathbf{R}^2$.*

(i) *They are linearly dependent if and only if $ad - bc = 0$.*

(ii) *If they are linearly independent, then they form a basis of $\mathbf{R}^2$.*

Proof. First work it out as an exercise (see Exercise 4). If you can't do it, you will find the proof in the answer section. It parallels closely the procedure of Example 4.

Let V be a vector space, and let $\{v_1,\dots,v_n)$ be a basis of V. The elements of V can be represented by n-tuples relative to this basis, as follows. If an element v of V is written as a linear combination

$$v = x_1 v_1 + \cdots + x_n v_n$$

of the basis elements, then we call $(x_1,\dots,x_n)$ the **coordinates** of v with respect to our basis, and we call x_i the i-th coordinate. The coordinates with respect to the usual basis $E_1,\dots,E_n$ of $\mathbf{R}^n$ are simply the coordinates as defined in Chapter I, §1.

The following theorem shows that there can only be one set of coordinates for a given vector.

Theorem 4.2. *Let V be a vector space. Let $v_1,\dots,v_n$ be linearly independent elements of V. Let $x_1,\dots,x_n$ and $y_1,\dots,y_n$ be numbers such that*

$$x_1 v_1 + \cdots + x_n v_n = y_1 v_1 + \cdots + y_n v_n.$$

Then we must have $x_i = y_i$ for all $i = 1,\dots,n$.

Proof. Subtract the right-hand side from the left-hand side. We get

$$x_1 v_1 - y_1 v_1 + \cdots + x_n v_n - y_n v_n = O.$$

We can write this relation also in the form

$$(x_1 - y_1)v_1 + \cdots + (x_n - y_n)v_n = O.$$

By definition, we must have $x_i - y_i = 0$ for all $i = 1,\dots,n$, thereby proving our assertion.

The theorem expresses the fact that when an element is written as a linear combination of $v_1,\dots,v_n$, then its coefficients $x_1,\dots,x_n$ are uniquely determined. This is true only when $v_1,\dots,v_n$ are linearly independent.

Example 5. Find the coordinates of $(1,0)$ with respect to the two vectors $(1,1)$ and $(-1,2)$.

We must find numbers a, b such that

$$a(1,1) + b(-1,2) = (1,0).$$

Writing this equation in terms of coordinates, we find

$$a - b = 1, \qquad a + 2b = 0.$$

Solving for a and b in the usual manner yields $b = -\frac{1}{3}$ and $a = \frac{2}{3}$. Hence the coordinates of $(1, 0)$ with respect to $(1, 1)$ and $(-1, 2)$ are $(\frac{2}{3}, -\frac{1}{3})$.

Example 6. The two functions e^t, e^{2t} are linearly independent. To prove this, suppose that there are numbers a, b such that

$$ae^t + be^{2t} = 0$$

(for all values of t). Differentiate this relation. We obtain

$$ae^t + 2be^{2t} = 0.$$

Subtract the first from the second relation. We obtain $be^{2t} = 0$, and hence $b = 0$. From the first relation, it follows that $ae^t = 0$, and hence $a = 0$. Hence e^t, e^{2t} are linearly independent.

Example 7. Let V be the vector space of all functions of a variable t. Let $f_1, \ldots, f_n$ be n functions. To say that they are linearly dependent is to say that there exist n numbers $a_1, \ldots, a_n$ not all equal to 0 such that

$$a_1 f_1(t) + \cdots + a_n f_n(t) = 0$$

for *all* values of t.

Warning. We emphasize that linear dependence for functions means that the above relation holds for all values of t. For instance, consider the relation

$$a \sin t + b \cos t = 0,$$

where a, b are two fixed numbers not both zero. There may be *some* values of t for which the above equation is satisfied. For instance, if $a \neq 0$ we then can solve

$$\frac{\sin t}{\cos t} = \frac{b}{a},$$

or in other words, $\tan t = b/a$ to get at least one solution. However, the above relation cannot hold for *all* values of t, and consequently $\sin t$, $\cos t$ are linearly independent, as functions.

Example 8. Let V be the vector space of functions generated by the two functions e^t, e^{2t}. Then the coordinates of the function

$$3e^t + 5e^{2t}$$

with respect to the basis $\{e^t, e^{2t}\}$ are $(3, 5)$.

When dealing with two vectors v, w there is another convenient way of expressing linear independence.

Theorem 4.3. *Let v, w be elements of a vector space V. They are linearly dependent if and only if one of them is a scalar multiple of the other, i.e. there is a number $c \neq 0$ such that we have $v = cw$ or $w = cv$.*

Proof. Left as an exercise, cf. Exercise 5.

In the light of this theorem, the condition imposed in various examples in the preceding section could be formulated in terms of two vectors being linearly independent.

Exercises III, §4

1. Show that the following vectors are linearly independent.
   (a) $(1, 1, 1)$ and $(0, 1, -2)$ (b) $(1, 0)$ and $(1, 1)$
   (c) $(-1, 1, 0)$ and $(0, 1, 2)$ (d) $(2, -1)$ and $(1, 0)$
   (e) $(\pi, 0)$ and $(0, 1)$ (f) $(1, 2)$ and $(1, 3)$
   (g) $(1, 1, 0)$, $(1, 1, 1)$,   (h) $(0, 1, 1)$, $(0, 2, 1)$,
       and $(0, 1, -1)$ and $(1, 5, 3)$

2. Express the given vector X as a linear combination of the given vectors A, B, and find the coordinates of X with respect to A, B.
   (a) $X = (1, 0)$, $A = (1, 1)$, $B = (0, 1)$
   (b) $X = (2, 1)$, $A = (1, -1)$, $B = (1, 1)$
   (c) $X = (1, 1)$, $A = (2, 1)$, $B = (-1, 0)$
   (d) $X = (4, 3)$, $A = (2, 1)$, $B = (-1, 0)$

3. Find the coordinates of the vector X with respect to the vectors A, B, C.
   (a) $X = (1, 0, 0)$, $A = (1, 1, 1)$, $B = (-1, 1, 0)$, $C = (1, 0, -1)$
   (b) $X = (1, 1, 1)$, $A = (0, 1, -1)$, $B = (1, 1, 0)$, $C = (1, 0, 2)$
   (c) $X = (0, 0, 1)$, $A = (1, 1, 1)$, $B = (-1, 1, 0)$, $C = (1, 0, -1)$

4. Let (a, b) and (c, d) be two vectors in $\mathbf{R}^2$.
   (i) If $ad - bc \neq 0$, show that they are linearly independent.
   (ii) If they are linearly independent, show that $ad - bc \neq 0$.
   (iii) If $ad - bc \neq 0$ show that they form a basis of $\mathbf{R}^2$.

5. (a) Let v, w be elements of a vector space. If v, w are linearly dependent, show that there is a number c such that $w = cv$, or $v = cw$.
   (b) Conversely, let v, w be elements of a vector space, and assume that there exists a number c such that $w = cv$. Show that v, w are linearly dependent.

6. Let $A_1, \ldots, A_r$ be vectors in $\mathbf{R}^n$, and assume that they are mutually perpendicular, in other words $A_i \perp A_j$ if $i \neq j$. Also assume that none of them is O. Prove that they are linearly independent.

7. Consider the vector space of all functions of a variable t. Show that the following pairs of functions are linearly independent.
   (a) $1, t$ (b) t, t^2 (c) t, t^4 (d) e^t, t (e) te^t, e^{2t} (f) $\sin t, \cos t$
   (g) $t, \sin t$ (h) $\sin t, \sin 2t$ (i) $\cos t, \cos 3t$

8. Consider the vector space of functions defined for $t > 0$. Show that the following pairs of functions are linearly independent.
   (a) $t, 1/t$ (b) $e^t, \log t$

9. What are the coordinates of the function $3 \sin t + 5 \cos t = f(t)$ with respect to the basis $\{\sin t, \cos t\}$?

10. Let D be the derivative d/dt. Let $f(t)$ be as in Exercise 9. What are the coordinates of the function $Df(t)$ with respect to the basis of Exercise 9?

In each of the following cases, exhibit a basis for the given space, and prove that it is a basis.

11. The space of 2×2 matrices.
12. The space of $m \times n$ matrices.
13. The space of $n \times n$ matrices all of whose components are 0 except possibly the diagonal components.
14. The **upper triangular matrices**, i.e. matrices of the following type:

$$\begin{pmatrix} a_{11} & a_{12} & \cdots & a_{1n} \\ 0 & a_{22} & \cdots & a_{2n} \\ \vdots & \vdots & & \vdots \\ 0 & 0 & \cdots & a_{nn} \end{pmatrix}.$$

15. (a) The space of symmetric 2×2 matrices.
    (b) The space of symmetric 3×3 matrices.
16. The space of symmetric $n \times n$ matrices.

III, §5. Dimension

We ask the question: Can we find three linearly independent elements in $\mathbf{R}^2$? For instance, are the elements

$$A = (1, 2), \qquad B = (-5, 7), \qquad C = (10, 4)$$

linearly independent? If you write down the linear equations expressing the relation

$$xA + yB + zC = 0,$$

you will find that you can solve them for x, y, z not equal to 0. Namely, these equations are:

$$x - 5y + 10z = 0,$$

$$2x + 7y + 4z = 0.$$

This is a system of two homogeneous equations in three unknowns, and we know by Theorem 2.1 of Chapter II that we can find a non-trivial solution (x, y, z) not all equal to zero. Hence A, B, C are linearly dependent.

We shall see in a moment that this is a general phenomenon. In $\mathbf{R}^n$, we cannot find more than n linearly independent vectors. Furthermore, we shall see that any n linearly independent elements of $\mathbf{R}^n$ must generate $\mathbf{R}^n$, and hence form a basis. Finally, we shall also see that if one basis of a vector space has n elements, and another basis has m elements, then $m = n$. In short, two bases must have the same number of elements. This property will allow us to define the **dimension** of a vector space as the number of elements in any basis. We now develop these ideas systematically.

Theorem 5.1. *Let V be a vector space, and let $\{v_1, \ldots, v_m\}$ generate V. Let $w_1, \ldots, w_n$ be elements of V and assume that $n > m$. Then $w_1, \ldots, w_n$ are linearly dependent.*

Proof. Since $\{v_1, \ldots, v_m\}$ generate V, there exist numbers (a_{ij}) such that we can write

$$w_1 = a_{11}v_1 + \cdots + a_{m1}v_m$$
$$\vdots \qquad \vdots \qquad\qquad \vdots$$
$$w_n = a_{1n}v_1 + \cdots + a_{mn}v_m.$$

If $x_1, \ldots, x_n$ are numbers, then

$$x_1 w_1 + \cdots + x_n w_n$$
$$= (x_1 a_{11} + \cdots + x_n a_{1n})v_1 + \cdots + (x_1 a_{m1} + \cdots + x_n a_{mn})v_m$$

(just add up the coefficients of $v_1, \ldots, v_m$ vertically downward). According to Theorem 2.1 of Chapter II, the system of equations

$$x_1 a_{11} + \cdots + x_n a_{1n} = 0$$
$$\vdots \qquad\qquad \vdots$$
$$x_1 a_{m1} + \cdots + x_n a_{mn} = 0$$

has a non-trivial solution, because $n > m$. In view of the preceding remark, such a solution $(x_1, \ldots, x_n)$ is such that

$$x_1 w_1 + \cdots + x_n w_n = 0.$$

as desired.

Theorem 5.2. *Let V be a vector space and suppose that one basis has n elements, and another basis has m elements. Then $m = n$.*

Proof. We apply Theorem 5.1 to the two bases. Theorem 5.1 implies that both alternatives $n > m$ and $m > n$ are impossible, and hence $m = n$.

Let V be a vector space having a basis consisting of n elements. We shall say that n is the **dimension** of V. If V consists of O alone, then V does not have a basis, and we shall say that V has dimension 0.

We may now reformulate the definitions of a line and a plane in an arbitrary vector space V. A **line passing through the origin** is simply a one-dimensional subspace. A **plane passing through the origin** is simply a two-dimensional subspace.

An arbitrary **line** is obtained as the translation of a one-dimensional subspace. An arbitrary **plane** is obtained as the translation of a two-dimensional subspace. When a basis $\{v_1\}$ has been selected for a one-dimensional space, then the points on a line are expressed in the usual form

$$P + t_1 v_1 \text{ with all possible numbers } t_1.$$

When a basis $\{v_1, v_2\}$ has been selected for a two-dimensional space, then the points on a plane are expressed in the form

$$P + t_1 v_1 + t_2 v_2 \text{ with possible numbers } t_1, t_2.$$

Let $\{v_1, \ldots, v_n\}$ be a set of elements of a vector space V. Let r be a positive integer $\leqq n$. We shall say that $\{v_1, \ldots, v_r\}$ is a **maximal** subset of linearly independent elements if $v_1, \ldots, v_r$ are linearly independent, and if in addition, given any v_i with $i > r$, the elements $v_1, \ldots, v_r, v_i$ are linearly dependent.

The next theorem gives us a useful criterion to determine when a set of elements of a vector space is a basis.

Theorem 5.3. *Let $\{v_1, \ldots, v_n\}$ be a set of generators of a vector space V. Let $\{v_1, \ldots, v_r\}$ be a maximal subset of linearly independent elements. Then $\{v_1, \ldots, v_r\}$ is a basis of V.*

Proof. We must prove that $v_1, \ldots, v_r$ generate V. We shall first prove that each v_i (for $i > r$) is a linear combination of $v_1, \ldots, v_r$. By hypothesis, given v_i, there exists numbers $x_1, \ldots, x_r, y$ not all 0 such that

$$x_1 v_1 + \cdots + x_r v_r + y v_i = O.$$

Furthermore, $y \neq 0$, because otherwise, we would have a relation of linear dependence for $v_1, \ldots, v_r$. Hence we can solve for v_i, namely

$$v_i = \frac{x_1}{-y} v_1 + \cdots + \frac{x_r}{-y} v_r,$$

thereby showing that v_i is a linear combination of $v_1, \ldots, v_r$.

Next, let v be any element of V. There exist numbers $c_1, \ldots, c_n$ such that

$$v = c_1 v_1 + \cdots + c_n v_n.$$

In this relation, we can replace each v_i $(i > r)$ by a linear combination of $v_1, \ldots, v_r$. If we do this, and then collect terms, we find that we have expressed v as a linear combination of $v_1, \ldots, v_r$. This proves that $v_1, \ldots, v_r$ generate V, and hence form a basis of V.

We shall now give criteria which allow us to tell when elements of a vector space constitute a basis.

Let $v_1, \ldots, v_n$ be linearly independent elements of a vector space V. We shall say that they form a **maximal set of linearly independent elements of** V if given any element w of V, the elements $w, v_1, \ldots, v_n$ are linearly dependent.

Theorem 5.4. *Let V be a vector space, and $\{v_1, \ldots, v_n\}$ a maximal set of linearly independent elements of V. Then $\{v_1, \ldots, v_n\}$ is a basis of V.*

Proof. We must now show that $v_1, \ldots, v_n$ generate V, i.e. that every element of V can be expressed as a linear combination of $v_1, \ldots, v_n$. Let w be an element of V. The elements $w, v_1, \ldots, v_n$ of V must be linearly dependent by hypothesis, and hence there exist numbers $x_0, x_1, \ldots, x_n$ not all 0 such that

$$x_0 w + x_1 v_1 + \cdots + x_n v_n = O.$$

We cannot have $x_0 = 0$, because if that were the case, we would obtain a relation of linear dependence among $v_1, \ldots, v_n$. Therefore we can solve for w in terms of $v_1, \ldots, v_n$, namely

$$w = -\frac{x_1}{x_0} v_1 - \cdots - \frac{x_n}{x_0} v_n.$$

This proves that w is a linear combination of $v_1, \ldots, v_n$, and hence that $\{v_1, \ldots, v_n\}$ is a basis.

Theorem 5.5. *Let V be a vector space of dimension n, and let $v_1, \ldots, v_n$ be linearly independent elements of V. Then $v_1, \ldots, v_n$ constitute a basis of V.*

Proof. According to Theorem 5.1., $\{v_1,\ldots,v_n\}$ is a maximal set of linearly independent elements of V. Hence it is a basis by Theorem 5.4.

Theorem 5.6. *Let V be a vector space of dimension n and let W be a subspace, also of dimension n. Then $W = V$.*

Proof. A basis for W must also be a basis for V.

Theorem 5.7. *Let V be a vector space of dimension n. Let r be a positive integer with $r < n$, and let $v_1,\ldots,v_r$ be linearly independent elements of V. Then one can find elements $v_{r+1},\ldots,v_n$ such that*

$$\{v_1,\ldots,v_n\}$$

is a basis of V.

Proof. Since $r < n$ we know that $\{v_1,\ldots,v_r\}$ cannot form a basis of V, and thus cannot be a maximal set of linearly independent elements of V. In particular, we can find v_{r+1} in V such that

$$v_1,\ldots,v_{r+1}$$

are linearly independent. If $r + 1 < n$, we can repeat the argument. We can thus proceed stepwise (by induction) until we obtain n linearly independent elements $\{v_1,\ldots,v_n\}$. These must be a basis by Theorem 5.4, and our corollary is proved.

Theorem 5.8. *Let V be a vector space having a basis consisting of n elements. Let W be a subspace which does not consist of O alone. Then W has a basis, and the dimension of W is $\leq n$.*

Proof. Let w_1 be a non-zero element of W. If $\{w_1\}$ is not a maximal set of linearly independent elements of W, we can find an element w_2 of W such that w_1, w_2 are linearly independent. Proceeding in this manner, one element at a time, there must be an integer $m \leq n$ such that we can find linearly independent elements $w_1, w_2,\ldots,w_m$, and such that

$$\{w_1,\ldots,w_m\}$$

is a maximal set of linearly independent elements of W (by Theorem 5.1 we cannot go on indefinitely finding linearly independent elements, and the number of such elements is at most n). If we now use Theorem 5.4, we conclude that $\{w_1,\ldots,w_m\}$ is a basis for W.

Exercises III, §5

1. What is the dimension of the following spaces (refer to Exercises 11 through 16 of the preceding section):
   (a) 2×2 matrices   (b) $m \times n$ matrices
   (c) $n \times n$ matrices all of whose components are 0 expect possibly on the diagonal.
   (d) Upper triangular $n \times n$ matrices.
   (e) Symmetric 2×2 matrices.
   (f) Symmetric 3×3 matrices.
   (g) Symmetric $n \times n$ matrices.

2. Let V be a subspace of $\mathbf{R}^2$. What are the possible dimensions for V? Show that if $V \neq \mathbf{R}^2$, then either $V = \{O\}$, or V is a straight line passing through the origin.

3. Let V be a subspace of $\mathbf{R}^3$. What are the possible dimensions for V? Show that if $V \neq \mathbf{R}^3$, then either $V = \{O\}$, or V is a straight line passing through the origin, or V is a plane passing through the origin.

III, §6. The Rank of a Matrix

Let

$$A = \begin{pmatrix} a_{11} & \cdots & a_{1n} \\ \vdots & & \vdots \\ a_{m1} & \cdots & a_{mn} \end{pmatrix}$$

be an $m \times n$ matrix. The columns of A generate a vector space, which is a subspace of $\mathbf{R}^m$. The dimension of that subspace is called the **column rank** of A. In light of Theorem 5.4, the column rank is equal to the maximum number of linearly independent columns. Similarly, the rows of A generate a subspace of $\mathbf{R}^n$, and the dimension of this subspace is called **the row rank**. Again by Theorem 5.4, the row rank is equal to the maximum number of linearly independent rows. We shall prove below that these two ranks are equal to each other. We shall give two proofs. The first in this section depends on certain operations on the rows and columns of a matrix. Later we shall give a more geometric proof using the notion of perpendicularity.

We define the **row space** of A to be the subspace generated by the rows of A. We define the **column space** of A to be the subspace generated by the columns.

Consider the following operations on the rows of a matrix.

Row 1. Adding a scalar multiple of one row to another.

Row 2. Interchanging rows.

Row 3. Multiplying one row by a non-zero scalar.

These are called the row operations (sometimes, the elementary row operations). We have similar operations for columns, which will be denoted by **Col 1, Col 2, Col 3** respectively. We shall study the effect of these operations on the ranks.

First observe that each one of the above operations has an inverse operation in the sense that by performing similar operations we can revert to the original matrix. For instance, let us change a matrix A by adding c times the second row to the first. We obtain a new matrix B whose rows are

$$B_1 = A_1 + cA_2, \ A_2, \dots, A_m.$$

If we now add $-cA_2$ to the first row of B, we get back A_1. A similar argument can be applied to any two rows.

If we interchange two rows, then interchange them again, we revert to the original matrix.

If we multiply a row by a number $c \neq 0$, then multiplying again by c^{-1} yields the original row.

Theorem 6.1. *Row and column operations do not change the row rank of a matrix, nor do they change the column rank.*

Proof. First we note that interchanging rows of a matrix does not affect the row rank since the subspace generated by the rows is the same, no matter in what order we take the rows.

Next, suppose we add a scalar multiple of one row to another. We keep the notation before the theorem, so the new rows are

$$B_1 = A_1 + cA_2, \ A_2, \dots, A_m.$$

Any linear combination of the rows of B, namely any linear combination of

$$B_1, \ A_2, \dots, A_m$$

is also a linear combination of $A_1, \ A_2, \dots, A_m$. Consequently the row space of B is contained in the row space of A. Hence by Theorem 5.6, we have

$$\text{row rank of } B \leq \text{row rank of } A.$$

Since A is also obtained from B by a similar operation, we get the reverse inequality

$$\text{row rank of } A \leq \text{row rank of } B.$$

Hence these two row ranks are equal.

Third, if we multiply a row A_i by $c \neq 0$, we get the new row cA_i. But $A_i = c^{-1}(cA_i)$, so the row spaces of the matrix A and the new matrix obtained by multiplying the row by c are the same. Hence the third operation also does not change the row rank.

We could have given the above argument with any pair of rows A_i, A_j ($i \neq j$), so we have seen that row operations do not change the row rank.

We now prove that they do not change the column rank.

Again consider the matrix obtained by adding a scalar multiple of the second row to the first:

$$B = \begin{pmatrix} a_{11} + ca_{21} & a_{12} + ca_{22} & \cdots & a_{1n} + ca_{2n} \\ a_{21} & a_{22} & \cdots & a_{2n} \\ \vdots & \vdots & & \vdots \\ a_{m1} & a_{m2} & \cdots & a_{mn} \end{pmatrix}.$$

Let $B^1, \ldots, B^n$ be the columns of this new matrix B. We shall see that the relation of linear dependence between the columns of B are precisely the same as the relations of linear dependence between the columns of A. In other words:

A vector $X = (x_1, \ldots, x_n)$ gives a relation of linear dependence

$$x_1 B^1 + \cdots + x_n B^n = O$$

between the columns of B if and only if X gives a relation of linear dependence

$$x_1 A^1 + \cdots + x_n A^n = O$$

between the columns of A.

Proof. We know from Chapter II, §2 that a relation of linear dependence among the columns can be written in terms of the dot product with the rows of the matrix. So suppose we have a relation

$$x_1 B^1 + \cdots + x_n B^n = O.$$

This is equivalent with the fact that

$$X \cdot B_i = 0 \qquad \text{for} \quad i = 1, \ldots, m.$$

Therefore

$$X \cdot (A_1 + cA_2) = 0, \quad X \cdot A_2 = 0, \quad \ldots, \quad X \cdot A_m = 0.$$

The first equation can be written

$$X \cdot A_1 + cX \cdot A_2 = 0.$$

Since $X \cdot A_2 = 0$ we conclude that $X \cdot A_1 = 0$. Hence X is perpendicular to the rows of A. Hence X gives a linear relation among the columns of A. The converse is proved similarly.

The above statement proves that if r among the columns of B are linearly independent, then r among the columns of A are also linearly independent, and conversely. Therefore A and B have the same column rank.

We leave the verification that the other row operations do not change the column ranks to the reader.

Similarly, one proves that the column operations do not change the row rank. The situation is symmetric between rows and columns. This concludes the proof of the theorem.

Theorem 6.2. *Let A be a matrix of row rank r. By a succession of row and column operations, the matrix can be transformed to the matrix having components equal to 1 on the diagonal of the first r rows and columns, and 0 everywhere else.*

$$
\left. \left\{ \begin{array}{c} r \\ \end{array} \right. \right.
\overbrace{}^{r}
\begin{pmatrix}
1 & 0 & \cdots & 0 & 0 & \cdots & 0 \\
0 & 1 & \cdots & 0 & 0 & \cdots & 0 \\
\vdots & \vdots & \ddots & \vdots & \vdots & & \vdots \\
0 & 0 & \cdots & 1 & 0 & \cdots & 0 \\
0 & 0 & \cdots & 0 & 0 & \cdots & 0 \\
\vdots & \vdots & & \vdots & \vdots & \ddots & \vdots \\
0 & 0 & \cdots & 0 & 0 & \cdots & 0
\end{pmatrix}.
$$

In particular, the row rank is equal to the column rank.

Proof. Suppose $r \neq 0$ so the matrix is not the zero matrix. Some component is not zero. After interchanging rows and columns, we may assume that this component is in the upper left-hand corner, that is this component is equal to $a_{11} \neq 0$. Now we go down the first column. We multiply the first row by a_{21}/a_{11} and subtract it from the second row. We then obtain a new matrix with 0 in the first place of the second row. Next we multiply the first row by a_{31}/a_{11} and subtract it from the third row. Then our new matrix has first component equal to 0 in the third row. Proceeding in the same way, we can transform the matrix so that it is of the form

$$
\begin{pmatrix}
a_{11} & a_{12} & \cdots & a_{1n} \\
0 & a_{22} & \cdots & a_{2n} \\
\vdots & \vdots & & \vdots \\
0 & a_{m2} & \cdots & a_{mn}
\end{pmatrix}.
$$

Next, we subtract appropriate multiples of the first column from the second, third, ..., n-th column to get zeros in the first row. This transforms the matrix to a matrix of type

$$\begin{pmatrix} a_{11} & 0 & \cdots & 0 \\ 0 & a_{22} & \cdots & a_{2n} \\ \vdots & \vdots & & \vdots \\ 0 & a_{m2} & \cdots & a_{mn} \end{pmatrix}.$$

Now we have an $(m-1) \times (n-1)$ matrix in the lower right. If we perform row and column operations on all but the first row and column, then first we do not disturb the first component a_{11}; and second we can repeat the argument, in order to obtain a matrix of the form

$$\begin{pmatrix} a_{11} & 0 & 0 & \cdots & 0 \\ 0 & a_{22} & 0 & \cdots & 0 \\ 0 & 0 & a_{33} & \cdots & a_{3n} \\ \vdots & \vdots & \vdots & & \vdots \\ 0 & 0 & a_{m3} & \cdots & a_{mn} \end{pmatrix}.$$

Proceeding stepwise by induction we reach a matrix of the form

$$\begin{pmatrix} a_{11} & 0 & \cdots & 0 & 0 \\ 0 & a_{22} & \cdots & 0 & \vdots \\ \vdots & \vdots & & \vdots & \vdots \\ 0 & 0 & \cdots & a_{ss} & \vdots \\ 0 & 0 & \cdots & 0 & 0 \end{pmatrix}$$

with diagonal elements $a_{11}, \ldots, a_{ss}$ which are $\neq 0$. We divide the first row by a_{11}, the second row by a_{22}, etc. We then obtain a matrix

$$\begin{pmatrix} 1 & 0 & \cdots & 0 & 0 \\ 0 & 1 & \cdots & 0 & \vdots \\ \vdots & \vdots & \ddots & \vdots & \vdots \\ 0 & 0 & \cdots & 1 & 0 \\ 0 & 0 & \cdots & 0 & 0 \end{pmatrix}.$$

Thus we have the unit $s \times s$ matrix in the upper left-hand corner, and zeros everywhere else. Since row and column operations do not change the row or column rank, it follows that $r = s$, and also that the row rank is equal to the column rank. This proves the theorem.

Since we have proved that the row rank is equal to the column rank, we can now omit "row" or "column" and just speak of the **rank** of a matrix. Thus by definition the **rank** of a matrix is equal to the dimension of the space generated by the rows.

Remark. Although the systematic procedure provides an effective method to find the rank, in practice one can usually take shortcuts to get as many zeros as possible by making row and column operations, so that at some point it becomes obvious what the rank of the matrix is.

Of course, one can also use the simple mechanism of linear equations to find the rank.

Example. Find the rank of the matrix

$$\begin{pmatrix} 2 & 1 & 1 \\ 0 & 1 & -1 \end{pmatrix}.$$

There are only two rows, so the rank is at most 2. On the other hand, the two columns

$$\begin{pmatrix} 2 \\ 0 \end{pmatrix} \quad \text{and} \quad \begin{pmatrix} 1 \\ 1 \end{pmatrix}$$

are linearly independent, for if a, b are numbers such that

$$a\begin{pmatrix} 2 \\ 0 \end{pmatrix} + b\begin{pmatrix} 1 \\ 1 \end{pmatrix} = \begin{pmatrix} 0 \\ 0 \end{pmatrix},$$

then

$$2a + b = 0,$$
$$b = 0,$$

so that $a = 0$. Therefore the two columns are linearly independent, and the rank is equal to 2.

Later we shall also see that determinants give a computation way of determining when vectors are linearly independent, and thus can be used to determine the rank.

Example. Find the rank of the matrix.

$$\begin{pmatrix} 1 & 2 & -3 \\ 2 & 1 & 0 \\ -2 & -1 & 3 \\ -1 & 4 & -2 \end{pmatrix}.$$

We subtract twice the first column from the second and add 3 times the first column to the third. This gives

$$\begin{pmatrix} 1 & 0 & 0 \\ 2 & -3 & 6 \\ -2 & 3 & -3 \\ -1 & 6 & -5 \end{pmatrix}.$$

We add 2 times the second column to the third. This gives

$$\begin{pmatrix} 1 & 0 & 0 \\ 2 & -3 & 0 \\ -2 & 3 & 3 \\ -1 & 6 & 7 \end{pmatrix}.$$

This matrix is in column echelon form, and it is immediate that the first three rows or columns are linearly independent. Since there are only three columns, it follows that the rank is 3.

Exercises III, §6

1. Find the rank of the following matrices.

(a) $\begin{pmatrix} 2 & 1 & 3 \\ 7 & 2 & 0 \end{pmatrix}$

(b) $\begin{pmatrix} -1 & 2 & -2 \\ 3 & 4 & -5 \end{pmatrix}$

(c) $\begin{pmatrix} 1 & 2 & 7 \\ 2 & 4 & -1 \end{pmatrix}$

(d) $\begin{pmatrix} 1 & 2 & -3 \\ -1 & -2 & 3 \\ 4 & 8 & -12 \\ 0 & 0 & 0 \end{pmatrix}$

(e) $\begin{pmatrix} 2 & 0 \\ 0 & -5 \end{pmatrix}$

(f) $\begin{pmatrix} -1 & 0 & 1 \\ 0 & 2 & 3 \\ 0 & 0 & 7 \end{pmatrix}$

(g) $\begin{pmatrix} 2 & 0 & 0 \\ -5 & 1 & 2 \\ 3 & 8 & -7 \end{pmatrix}$

(h) $\begin{pmatrix} 1 & 2 & -3 \\ -1 & -2 & 3 \\ 4 & 8 & -12 \\ 1 & -1 & 5 \end{pmatrix}$

(i) $\begin{pmatrix} 1 & 1 & 0 & 1 \\ 1 & 2 & 2 & 1 \\ 3 & 4 & 2 & 3 \end{pmatrix}$

2. Let A be a triangular matrix

$$\begin{pmatrix} a_{11} & a_{12} & \cdots & a_{1n} \\ 0 & a_{22} & \cdots & a_{2n} \\ \vdots & \vdots & & \vdots \\ 0 & 0 & \cdots & a_{nn} \end{pmatrix}$$

Assume that none of the diagonal elements is equal to 0. What is the rank of A?

3. Let A be an $m \times n$ matrix and let B be an $n \times r$ matrix, so we can form the product AB.

(a) Show that the columns of AB are linear combinations of the columns of A. Thus prove that

$$\text{rank } AB \leq \text{rank } A.$$

(b) Prove that rank $AB \leq$ rank B. [*Hint*: Use the fact that

$$\text{rank } AB = \text{rank } {}^t(AB)$$

and

$$\text{rank } B = \text{rank } {}^tB.]$$

CHAPTER IV

Linear Mappings

We shall first define the general notion of a mapping, which generalizes the notion of a function. Among mappings, the linear mappings are the most important. A good deal of mathematics is devoted to reducing questions concerning arbitrary mappings to linear mappings. For one thing, they are interesting in themselves, and many mappings are linear. On the other hand, it is often possible to approximate an arbitrary mapping by a linear one, whose study is much easier than the study of the original mapping. This is done in the calculus of several variables.

IV, §1. Mappings

Let S, S' be two sets. A **mapping** from S to S' is an association which to every element of S associates an element of S'. Instead of saying that F is a mapping from S into S', we shall often write the symbols $F: S \to S'$. A mapping will also be called a **map**, for the sake of brevity.

A function is a special type of mapping, namely it is a mapping from a set into the set of numbers, i.e. into **R**.

We extend to mappings some of the terminology we have used for functions. For instance, if $T: S \to S'$ is a mapping, and if u is an element of S, then we denote by $T(u)$, or Tu, the element of S' associated to u by T. We call $T(u)$ the **value** of T at u, or also the **image** of u under T. The symbols $T(u)$ are read "T of u". The set of all elements $T(u)$, when u ranges over all elements of S, is called the **image** of T. If W is a subset of S, then the set of elements $T(w)$, when w ranges over all elements of W, is called the **image** of W under T, and is denoted by $T(W)$.

Let $F: S \to S'$ be a map from a set S into a set S'. If x is an element of S, we often write

$$x \mapsto F(x)$$

with a special arrow $\mapsto$ to denote the image of x under F. Thus, for instance, we would speak of the map F such that $F(x) = x^2$ as the map $x \mapsto x^2$.

Example 1. For any set S we have the identity mapping $I: S \to S$. It is defined by $I(x) = x$ for all x.

Example 2. Let S and S' be both equal to **R**. Let $f: \mathbf{R} \to \mathbf{R}$ be the function $f(x) = x^2$ (i.e. the function whose value at a number x is x^2). Then f is a mapping from **R** into **R**. Its image is the set of numbers ≥ 0.

Example 3. Let S be the set of numbers ≥ 0, and let $S' = \mathbf{R}$. Let $g: S \to S'$ be the function such that $g(x) = x^{1/2}$. Then g is a mapping from S into **R**.

Example 4. Let S be the set of functions having derivatives of all orders on the interval $0 < t < 1$, and let $S' = S$. Then the derivative $D = d/dt$ is a mapping from S into S. Indeed, our map D associates the function $df/dt = Df$ to the function f. According to our terminology, Df is the value of the mapping D at f.

Example 5. Let S be the set $\mathbf{R}^3$, i.e. the set of 3-tuples. Let $A = (2, 3, -1)$. Let $L: \mathbf{R}^3 \to \mathbf{R}$ be the mapping whose value at a vector $X = (x, y, z)$ is $A \cdot X$. Then $L(X) = A \cdot X$. If $X = (1, 1, -1)$, then the value of L at X is 6.

Just as we did with functions, we describe a mapping by giving its values. Thus, instead of making the statement in Example 5 describing the mapping L, we would also say: Let $L: \mathbf{R}^3 \to \mathbf{R}$ be the mapping $L(X) = A \cdot X$. This is somewhat incorrect, but is briefer, and does not usually give rise to confusion. More correctly, we can write $X \mapsto L(X)$ or $X \mapsto A \cdot X$ with the special arrow $\mapsto$ to denote the effect of the map L on the element X.

Example 6. Let $F: \mathbf{R}^2 \to \mathbf{R}^2$ be the mapping given by

$$F(x, y) = (2x, 2y).$$

Describe the image under F of the points lying on the circle $x^2 + y^2 = 1$.

Let (x, y) be a point on the circle of radius 1.
Let $u = 2x$ and $v = 2y$. Then u, v satisfy the relation

$$(u/2)^2 + (v/2)^2 = 1$$

or in other words,

$$\frac{u^2}{4} + \frac{v^2}{4} = 1.$$

Hence (u, v) is a point on the circle of radius 2. Therefore the image under F of the circle of radius 1 is a subset of the circle of radius 2. Conversely, given a point (u, v) such that

$$u^2 + v^2 = 4,$$

let $x = u/2$ and $y = v/2$. Then the point (x, y) satisfies the equation

$$x^2 + y^2 = 1,$$

and hence is a point on the circle of radius 1. Furthermore,

$$F(x, y) = (u, v).$$

Hence every point on the circle of radius 2 is the image of some point on the circle of radius 1. We conclude finally that the image of the circle of radius 1 under F is precisely the circle of radius 2.

Note. In general, let S, S' be two sets. To prove that $S = S'$, one frequently proves that S is a subset of S' and that S' is a subset of S. This is what we did in the preceding argument.

Example 7. This example is particularly important in geometric applications. Let V be a vector space, and let u be a fixed element of V. We let

$$T_u: V \to V$$

be the map such that $T_u(v) = v + u$. We call T_u the **translation** by u. If S is any subset of V, then $T_u(S)$ is called the **translation of S by** u, and consists of all vectors $v + u$, with $v \in S$. We often denote it by $S + u$. In the next picture, we draw a set S and its translation by a vector u.

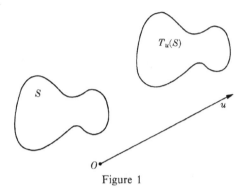

Figure 1

Example 8. Rotation counterclockwise around the origin by an angle θ is a mapping, which we may denote by R_θ. Let $\theta = \pi/2$. The image of the point $(1, 0)$ under the rotation $R_{\pi/2}$ is the point $(0, 1)$. We may write this as

$$R_{\pi/2}(1, 0) = (0, 1).$$

Example 9. Let S be a set. A mapping from S into $\mathbf{R}$ will be called a **function**, and the set of such functions will be called the set of functions defined on S. Let f, g be two functions defined on S. We can define their sum just as we did for functions of numbers, namely $f + g$ is the function whose value at an element t of S is $f(t) + g(t)$. We can also define the product of f by a number c. It is the function whose value at t is $cf(t)$. Then the set of mappings from S into $\mathbf{R}$ is a vector space.

Example 10. Let S be a set and let V be a vector space. Let F, G be two mappings from S into V. We can define their **sum** in the same way as we defined the sum of functions, namely the sum $F + G$ is the mapping whose value at an element t of S is $F(t) + G(t)$. We also define the **product** of F by a number c to be the mapping whose value at an element t of S is $cF(t)$. It is easy to verify that conditions **VS 1** through **VS 8** are satisfied.

Exercises IV, §1

1. In Example 4, give Df as a function of x when f is the function:
   (a) $f(x) = \sin x$ (b) $f(x) = e^x$ (c) $f(x) = \log x$

2. Let $P = (0, 1)$. Let R be rotation by $\pi/4$. Give the coordinates of the image of P under R, i.e. give $R(P)$.

3. In Example 5, give $L(X)$ when X is the vector:
   (a) $(1, 2, -3)$ (b) $(-1, 5, 0)$ (c) $(2, 1, 1)$

4. Let $F: \mathbf{R} \to \mathbf{R}^2$ be the mapping such that $F(t) = (e^t, t)$. What is $F(1)$, $F(0)$, $F(-1)$?

5. Let $G: \mathbf{R} \to \mathbf{R}^2$ be the mapping such that $G(t) = (t, 2t)$. Let F be as in Exercise 4. What is $(F + G)(1)$, $(F + G)(2)$, $(F + G)(0)$?

6. Let F be as in Exercise 4. What is $(2F)(0)$, $(\pi F)(1)$?

7. Let $A = (1, 1, -1, 3)$. Let $F: \mathbf{R}^4 \to \mathbf{R}$ be the mapping such that for any vector $X = (x_1, x_2, x_3, x_4)$ we have $F(X) = X \cdot A + 2$. What is the value of $F(X)$ when (a) $X = (1, 1, 0, -1)$ and (b) $X = (2, 3, -1, 1)$?

In Exercises 8 through 12, refer to Example 6. In each case, to prove that the image is equal to a certain set S, you must prove that the image is contained in S, and also that every element of S is in the image.

8. Let $F: \mathbf{R}^2 \to \mathbf{R}^2$ be the mapping defined by $F(x, y) = (2x, 3y)$. Describe the image of the points lying on the circle $x^2 + y^2 = 1$.

9. Let $F: \mathbf{R}^2 \to \mathbf{R}^2$ be the mapping defined by $F(x, y) = (xy, y)$. Describe the image under F of the straight line $x = 2$.

10. Let F be the mapping defined by $F(x, y) = (e^x \cos y, e^x \sin y)$. Describe the image under F of the line $x = 1$. Describe more generally the image under F of a line $x = c$, where c is a constant.

11. Let F be the mapping defined by $F(t, u) = (\cos t, \sin t, u)$. Describe geometrically the image of the (t, u)-plane under F.

12. Let F be the mapping defined by $F(x, y) = (x/3, y/4)$. What is the image under F of the ellipse

$$\frac{x^2}{9} + \frac{y^2}{16} = 1?$$

IV, §2. Linear Mappings

Let V, W be two vector spaces. A **linear mapping**

$$L: V \to W$$

is a mapping which satisfies the following two properties. First, for any elements u, v in V, and any scalar c, we have:

LM 1. $L(u + v) = L(u) + L(v)$.

LM 2. $L(cu) = cL(u)$.

Example 1. The most important linear mapping of this course is described as follows. Let A be a given $m \times n$ matrix. Define

$$L_A: \mathbf{R}^n \to \mathbf{R}^m$$

by the formula

$$L_A(X) = AX.$$

Then L_A is **linear**. Indeed, this is nothing but a summary way of expressing the properties

$$A(X + Y) = AX + AY \quad \text{and} \quad A(cX) = cAX$$

for any *vertical* X, Y in $\mathbf{R}^n$ and any number c.

Example 2. The **dot product** is essentially a special case of the first
example. Let $A = (a_1, \ldots, a_n)$ be a fixed vector, and define

$$L_A(X) = A \cdot X.$$

Then L_A is a linear map from $\mathbf{R}^n$ into $\mathbf{R}$, because

$$A \cdot (X + Y) = A \cdot X + A \cdot Y \quad \text{and} \quad A \cdot (cX) = c(A \cdot X).$$

Note that the dot product can also be viewed as multiplication of matrices if we view A as a row vector, and X as a *column* vector.

Example 3. Let V be any vector space. The mapping which associates
to any element u of V this element itself is obviously a linear mapping,
which is called the **identity** mapping. We denote it by I. Thus $I(u) = u$.

Example 4. Let V, W be any vector spaces. The mapping which associates the element O in W to any element u of V is called the **zero**
mapping and is obviously linear.

Example 5. Let V be the set of functions which have derivatives of all
orders. Then the derivative $D: V \to V$ is a linear mapping. This is simply
a brief way of summarizing standard properties of the derivative, namely.

$$D(f + g) = Df + Dg,$$
$$D(cf) = cD(f).$$

Example 6. Let $V = \mathbf{R}^3$ be the vector space of vectors in 3-space. Let
$V' = \mathbf{R}^2$ be the vector space of vectors in 2-space. We can define a
mapping.

$$F: \mathbf{R}^3 \to \mathbf{R}^2$$

by the projection, namely $F(x, y, z) = (x, y)$. We leave it to you to check
that the conditions **LM 1** and **LM 2** are satisfied.

More generally, suppose $n = r + s$ is expressed as a sum of two positive integers. We can separate the coordinates $(x_1, \ldots, x_n)$ into two

bunches $(x_1, \ldots, x_r, x_{r+1}, \ldots, x_{r+s})$, namely the first r coordinates, and the last s coordinates. Let

$$F: \mathbf{R}^n \to \mathbf{R}^r$$

be the map such that $F(x_1, \ldots, x_n) = (x_1, \ldots, x_r)$. Then you can verify easily that F is linear. We call F the **projection on the first r coordinates**. Similarly, we would have a **projection on the last s coordinates**, by means of the linear map L such that

$$L(x_1, \ldots, x_n) = (x_{r+1}, \ldots, x_n).$$

Example 7. In the calculus of several variables, one defines the gradient of a function f to be

$$\operatorname{grad} f(X) = \left(\frac{\partial f}{\partial x_1}, \ldots, \frac{\partial f}{\partial x_n} \right).$$

Then for two functions f, g, we have

$$\operatorname{grad}(f + g) = \operatorname{grad} f + \operatorname{grad} g$$

and for any number c,

$$\operatorname{grad}(cf) = c \cdot \operatorname{grad} f.$$

Thus grad is a linear map.

Let $L: V \to W$ be a linear mapping. Let u, v, w be elements of V. Then

$$L(u + v + w) = L(u) + L(v) + L(w).$$

This can be seen stepwise, using the definition of linear mappings. Thus

$$L(u + v + w) = L(u + v) + L(w) = L(u) + L(v) + L(w).$$

Similarly, given a sum of more than three elements, an analogous property is satisfied. For instance, let $u_1, \ldots, u_n$ be elements of V. Then

$$L(u_1 + \cdots + u_n) = L(u_1) + \cdots + L(u_n).$$

The sum on the right can be taken in any order. A formal proof can easily be given by induction, and we omit it.

If $a_1, \ldots, a_n$ are numbers, then

$$L(a_1 u_1 + \cdots + a_n u_n) = a_1 L(u_1) + \cdots + a_n L(u_n).$$

We show this for three elements.

$$L(a_1 u + a_2 v + a_3 w) = L(a_1 u) + L(a_2 v) + L(a_3 w)$$
$$= a_1 L(u) + a_2 L(v) + a_3 L(w).$$

With the notation of summation signs, we would write

$$L\left(\sum_{i=1}^{n} a_i u_i\right) = \sum_{i=1}^{n} a_i L(u_i).$$

In practice, the following properties will be obviously satisfied, but it turns out they can be proved from the axioms of linear maps and vector spaces.

LM 3. *Let $L: V \to W$ be a linear map. Then $L(O) = O$.*

Proof. We have

$$L(O) = L(O + O) = L(O) + L(O).$$

Subtracting $L(O)$ from both sides yields $O = L(O)$, as desired.

LM 4. *Let $L: V \to W$ be a linear map. Then $L(-v) = -L(v)$.*

Proof. We have

$$O = L(O) = L(v - v) = L(v) + L(-v).$$

Add $-L(v)$ to both sides to get the desired assertion.

We observe that the values of a linear map are determined by knowing the values on the elements of a basis.

Example 8. Let $L: \mathbf{R}^2 \to \mathbf{R}^2$ be a linear map. Suppose that

$$L(1, 1) = (1, 4) \qquad \text{and} \qquad L(2, -1) = (-2, 3).$$

Find $L(3, -1)$.

To do this, we write $(3, -1)$ as a linear combination of $(1, 1)$ and $(2, -1)$. Thus we have to solve

$$(3, -1) = x(1, 1) + y(2, -1).$$

This amounts to solving

$$x + 2y = 3,$$
$$x - y = -1.$$

The solution is $x = \frac{1}{3}$, $y = \frac{4}{3}$. Hence

$$L(3, -1) = xL(1, 1) + yL(2, -1) = \frac{1}{3}(1, 4) + \frac{4}{3}(-2, 3) = \left(\frac{-7}{3}, \frac{16}{3}\right).$$

Example 9. Let V be a vector space, and let $L: V \to \mathbf{R}$ be a linear map. We contend that the set S of all elements v in V such that $L(v) < 0$ is convex.

Proof. Let $L(v) < 0$ and $L(w) < 0$. Let $0 < t < 1$. Then

$$L(tv + (1 - t)w) = tL(v) + (1 - t)L(w).$$

Then $tL(v) < 0$ and $(1 - t)L(w) < 0$ so $tL(v) + (1 - t)L(w) < 0$, whence $tv + (1 - t)w$ lies in S. If $t = 0$ or $t = 1$, then $tv + (1 - t)w$ is equal to v or w and this also lies in S. This proves our assertion.

For a generalization of this example, see Exercise 14.

The coordinates of a linear map

Let first

$$F: V \to \mathbf{R}^n$$

be any mapping. Then each value $F(v)$ is an element of $\mathbf{R}^n$, and so has coordinates. Thus we can write

$$F(v) = (F_1(v), \ldots, F_n(v)), \qquad \text{or} \qquad F = (F_1, \ldots, F_n).$$

Each F_i is a function of V into $\mathbf{R}$, which we write

$$F_i: V \to \mathbf{R}.$$

Example 10. Let $F: \mathbf{R}^2 \to \mathbf{R}^3$ be the mapping

$$F(x, y) = (2x - y, 3x + 4y, x - 5y).$$

Then

$$F_1(x, y) = 2x - y, \qquad F_2(x, y) = 3x + 4y, \qquad F_3(x, y) = x - 5y.$$

Observe that each coordinate function can be expressed in terms of a dot product. For instance, let

$$A_1 = (2, -1), \qquad A_2 = (3, 4), \qquad A_3 = (1, -5).$$

Then

$$F_i(x, y) = A_i \cdot (x, y) \qquad \text{for} \qquad i = 1, 2, 3.$$

Each function

$$X \mapsto A_i \cdot X$$

is linear. Quite generally:

Proposition 2.1. *Let* $F: V \to \mathbf{R}^n$ *be a mapping of a vector space* V *into* $\mathbf{R}^n$. *Then* F *is linear if and only if each coordinate function* $F_i: V \to \mathbf{R}$ *is linear, for* $i = 1, \ldots, n$.

Proof. For $v, w \in V$ we have

$$F(v + w) = (F_1(v + w), \ldots, F_n(v + w)),$$

$$F(v) = (F_1(v), \ldots, F_n(v)),$$

$$F(w) = (F_1(w), \ldots, F_n(w)).$$

Thus $F(v + w) = F(v) + F(w)$ if and only if $F_i(v + w) = F_i(v) + F_i(w)$ for all $i = 1, \ldots, n$ by the definition of addition of n-tuples. The same argument shows that if $c \in \mathbf{R}$, then $F(cv) = cF(v)$ if and only if

$$F_i(cv) = cF_i(v) \qquad \text{for all} \quad i = 1, \ldots, n.$$

This proves the proposition.

Example 10 (continued). The mapping of Example 10 is linear because each coordinate function is linear. Actually, if you write the vector (x, y) vertically, you should realize that the mapping F is in fact equal to L_A for some matrix A. What is this matrix A?

The vector space of linear maps

Let V, W be two vector spaces. We consider the set of all linear mappings from V into W, and denote this set by $\mathcal{L}(V, W)$, or simply $\mathcal{L}$ if the reference to V and W is clear. We shall define the addition of linear mappings and their multiplication by numbers in such a way as to make $\mathcal{L}$ into a vector space.

Let $L: V \to W$ and let $F: V \to W$ be two linear mappings. We define their **sum** $L + F$ to be the map whose value at an element u of V is $L(u) + F(u)$. Thus we may write

$$(L + F)(u) = L(u) + F(u).$$

The map $L + F$ is then a linear map. Indeed, it is easy to verify that the two conditions which define a linear map are satisfied. For any elements u, v of V, we have

$$(L + F)(u + v) = L(u + v) + F(u + v)$$

$$= L(u) + L(v) + F(u) + F(v)$$

$$= L(u) + F(u) + L((v) + F(v)$$

$$= (L + F)(u) + (L + F)(v).$$

Furthermore, if c is a number, then

$$(L + F)(cu) = L(cu) + F(cu)$$

$$= cL(u) + cF(u)$$

$$= c[L(u) + F(u)]$$

$$= c[(L + F)(u)].$$

Hence $L + F$ is a linear map.

If a is a number, and $L: V \to W$ is a linear map, we define a map aL from V into W by giving its value at an element u of V, namely $(aL)(u) = aL(u)$. Then it is easily verified that aL is a linear map. We leave this as an exercise.

We have just defined operations of addition and multiplication by numbers in our set $\mathscr{L}$. Furthermore, if $L: V \to W$ is a linear map, i.e. an element of $\mathscr{L}$, then we can define $-L$ to be $(-1)L$, i.e. the product of the number -1 by L. Finally, we have the **zero-map**, which to every element of V associates the element O of W. Then $\mathscr{L}$ is a vector space. In other words, the set of linear maps from V into W is itself a vector space. The verification that the rules **VS 1** through **VS 8** for a vector space are satisfied is easy and is left to the reader.

Example 11. Let $V = W$ be the vector space of functions which have derivatives of all orders. Let D be the derivative, and let I be the identity. If f is in V, then

$$(D + I)f = Df + f.$$

Thus, when $f(x) = e^x$, then $(D + I)f$ is the function whose value at x is $e^x + e^x = 2e^x$.

If $f(x) = \sin x$, then $(D + 3I)f$ is the function such that

$$((D + 3I)f)(x) = (Df)(x) + 3If(x) = \cos x + 3 \sin x.$$

We note that $3 \cdot I$ is a linear map, whose value at f is $3f$. Thus $(D + 3 \cdot I)f = Df + 3f$. At any number x, the value of $(D + 3 \cdot I)f$ is $Df(x) + 3f(x)$. We can also write $(D + 3I)f = Df + 3f$.

Exercises IV, §2

1. Determine which of the following mappings F are linear.
   (a) $F: \mathbf{R}^3 \to \mathbf{R}^2$ defined by $F(x, y, z) = (x, z)$. _yes_
   (b) $F: \mathbf{R}^4 \to \mathbf{R}^4$ defined by $F(X) = -X$.
   (c) $F: \mathbf{R}^3 \to \mathbf{R}^3$ defined by $F(X) = X + (0, -1, 0)$. _No_
   (d) $F: \mathbf{R}^2 \to \mathbf{R}^2$ defined by $F(x, y) = (2x + y, y)$.
   (e) $F: \mathbf{R}^2 \to \mathbf{R}^2$ defined by $F(x, y) = (2x, y - x)$. _yes_
   (f) $F: \mathbf{R}^2 \to \mathbf{R}^2$ defined by $F(x, y) = (y, x)$.
   (g) $F: \mathbf{R}^2 \to \mathbf{R}$ defined by $F(x, y) = xy$. _no_

2. Which of the mappings in Exercises 4, 7, 8, 9, of §1 are linear?

3. Let V, W be two vector spaces and let $F: V \to W$ be a linear map. Let U be the subset of V consisting of all elements v such that $F(v) = O$. Prove that U is a subspace of V.

4. Let $L: V \to W$ be a linear map. Prove that the image of L is a subspace of W. [This will be done in the next section, but try it now to give you practice.]

5. Let A, B be two $m \times n$ matrices. Assume that

$$AX = BX$$

for all n-tuples X. Show that $A = B$. This can also be stated in the form: If $L_A = L_B$ then $A = B$.

6. Let $T_u: V \to V$ be the translation by a vector u. For which vectors u is T_u a linear map?

7. Let $L: V \to W$ be a linear map.
   (a) If S is a line in V, show that the image $L(S)$ is either a line in W or a point.
   (b) If S is a line segment in V, between the points P and Q, show that the image $L(S)$ is either a point or a line segment in W. Between which points in W?
   (c) Let v_1, v_2 be linearly independent elements of V. Assume that $L(v_1)$ and $L(v_2)$ are linearly independent in W. Let P be an element of V, and let S be the parallelogram

$$P + t_1 v_1 + t_2 v_2 \quad \text{with} \quad 0 \le t_i \le 1 \quad \text{for} \quad i = 1, 2.$$

   Show that the image $L(S)$ is a parallelogram in W.
   (d) Let v, w be linearly independent elements of a vector space V. Let $F: V \to W$ be a linear map. Assume that $F(v)$, $F(w)$ are linearly depen-

dent. Show that the image under F of the parallelogram spanned by v and w is either a point or a line segment.

8. Let $E_1 = (1, 0)$ and $E_2 = (0, 1)$ as usual. Let F be a linear map from $\mathbf{R}^2$ into itself such that

$$F(E_1) = (1, 1) \quad \text{and} \quad F(E_2) = (-1, 2).$$

Let S be the square whose corners are at $(0, 0)$, $(1, 0)$, $(1, 1)$, and $(0, 1)$. Show that the image of this square under F is a parallelogram.

9. Let A, B be two non-zero vectors in the plane such that there is no constant $c \neq 0$ such that $B = cA$. Let L be a linear mapping of the plane into itself such that $L(E_1) = A$ and $L(E_2) = B$. Describe the image under L of the rectangle whose corners are $(0, 1)$, $(3, 0)$, $(0, 0)$, and $(3, 1)$.

10. Let $L: \mathbf{R}^2 \to \mathbf{R}^2$ be a linear map, having the following effect on the indicated vectors:
    (a) $L(3, 1) = (1, 2)$ and $L(-1, 0) = (1, 1)$
    (b) $L(4, 1) = (1, 1)$ and $L(1, 1) = (3, -2)$
    (c) $L(1, 1) = (2, 1)$ and $L(-1, 1) = (6, 3)$.
    In each case compute $L(1, 0)$.

11. Let L be as in (a), (b), (c), of Exercise 10. Find $L(0, 1)$.

12. Let V, W be two vector spaces, and $F: V \to W$ a linear map. Let $w_1, \ldots, w_n$ be elements of W which are linearly independent, and let $v_1, \ldots, v_n$ be elements of V such that $F(v_i) = w_i$ for $i = 1, \ldots, n$. Show that $v_1, \ldots, v_n$ are linearly independent.

13. (a) Let V be a vector space and $F: V \to \mathbf{R}$ a linear map. Let W be the subset of V consisting of all elements v such that $F(v) = 0$. Assume that $W \neq V$, and let v_0 be an element of V which does not lie in W. Show that every element of V can be written as a sum $w + cv_0$, with some w in W and some number c.
    (b) Show that W is a subspace of V. Let $\{v_1, \ldots, v_n\}$ be a basis of W. Show that $\{v_0, v_1, \ldots, v_n\}$ is a basis of V.

Convex sets

14. Show that the image of a convex set under a linear map is convex.

15. Let $L: V \to W$ be a linear map. Let T be a convex set in W and let S be the set of elements $v \in V$ such that $L(v) \in T$. Show that S is convex.

   Remark. Why do these exercises give a more general proof of what you should already have worked out previously? For instance: Let $A \in \mathbf{R}^n$ and let c be a number. Then the set of all $X \in \mathbf{R}^n$ such that $X \cdot A \geq c$ is convex. Also if S is a convex set and c is a number, then cS is convex. How do these statements fit as special cases of Exercises 14 and 15?

16. Let S be a convex set in V and let $u \in V$. Let $T_u: V \to V$ be the translation by u. Show that the image $T_u(S)$ is convex.

Eigenvectors and eigenvalues. Let V be a vector space, and let $L: V \to V$ be a linear map. An **eigenvector** v for L is an element of V such that there exists a scalar c with the property.

$$L(v) = cv.$$

The scalar c is called an **eigenvalue** of v with respect to L. If $v \neq 0$ then c is uniquely determined. When V is a vector space whose elements are functions, then an eigenvector is also called an **eigenfunction**.

17. (a) Let V be the space of differentiable functions on **R**. Let $f(t) = e^{ct}$, where c is some number. Let L be the derivative d/dt. Show that f is an eigenfunction for L. What is the eigenvalue?
    (b) Let L be the second derivative, that is

$$L(f) = \frac{d^2 f}{dt^2}$$

    for any function f. Show that the functions $\sin t$ and $\cos t$ are eigenfunctions of L. What are the eigenvalues?

18. Let $L: V \to V$ be a linear map, and let W be the subset of elements of V consisting of all eigenvectors of L with a given eigenvalue c. Show that W is a subspace.

19. Let $L: V \to V$ be a linear map. Let $v_1, \ldots, v_n$ be non-zero eigenvectors for L, with eigenvalues $c_1, \ldots, c_n$ respectively. Assume that $c_1, \ldots, c_n$ are distinct. Prove that $v_1, \ldots, v_n$ are linearly independent. [*Hint*: Use induction.]

IV, §3. The Kernel and Image of a Linear Map

Let $F: V \to W$ be a linear map. The **image** of F is the set of elements w in W such that there exists an element v of V such that $F(v) = w$.

The image of F is a subspace of W.

Proof. Observe first that $F(O) = O$, and hence O is in the image. Next, suppose that w_1, w_2 are in the image. Then there exist elements v_1, v_2 of V such that $F(v_1) = w_1$ and $F(v_2) = w_2$. Hence

$$F(v_1 + v_2) = F(v_1) + F(v_2) = w_1 + w_2,$$

thereby proving that $w_1 + w_2$ is in the image. If c is a number, then

$$F(cv_1) = cF(v_1) = cw_1.$$

Hence cw_1 is in the image. This proves that the image is a subspace of W.

Let V, W be vector spaces, and let $F: V \to W$ be a linear map. The set of elements $v \in V$ such that $F(v) = O$ is called the **kernel** of F.

The kernel of F is a subspace of V.

Proof. Since $F(O) = O$, we see that O is in the kernel. Let v, w be in the kernel. Then $F(v + w) = F(v) + F(w) = O + O = O$, so that $v + w$ is in the kernel. If c is a number, then $F(cv) = cF(v) = O$ so that cv is also in the kernel. Hence the kernel is a subspace.

Example 1. Let $L: \mathbf{R}^3 \to \mathbf{R}$ be the map such that

$$L(x, y, z) = 3x - 2y + z.$$

Thus if $A = (3, -2, 1)$, then we can write

$$L(X) = X \cdot A = A \cdot X.$$

Then the kernel of L is the set of solutions of the equation.

$$3x - 2y + z = 0.$$

Of course, this generalizes to n-space. If A is an arbitrary vector in $\mathbf{R}^n$, we can define the linear map

$$L_A: \mathbf{R}^n \to \mathbf{R}$$

such that $L_A(X) = A \cdot X$. Its kernel can be interpreted as the set of all X which are perpendicular to A.

Example 2. Let $P: \mathbf{R}^3 \to \mathbf{R}^2$ be the projection, such that

$$P(x, y, z) = (x, y).$$

Then P is a linear map whose kernel consists of all vectors in $\mathbf{R}^3$ whose first two coordinates are equal to 0, i.e. all vectors

$$(0, 0, z)$$

with arbitrary component z.

Example 3. Let A be an $m \times n$ matrix, and let

$$L_A: \mathbf{R}^n \to \mathbf{R}^m$$

be the linear map such that $L_A(X) = AX$. Then the kernel of L_A is precisely the subspace of solutions X of the linear equations

$$AX = O.$$

Example 4. Differential equations. Let D be the derivative. If the real variable is denoted by x, then we may also write $D = d/dx$. The derivative may be iterated, so the second derivative is denoted by D^2 (or $(d/dx)^2$). When applied to a function, we write D^2f, so that

$$(D^2f)(x) = \frac{d^2f}{dx^2}.$$

Similarly for D^3, D^4,...,D^n for the n-th derivative.

Now let V be the vector space of functions which admit derivatives of all orders. Let $a_1, \ldots, a_m$ be numbers, and let g be an element of V, that is an infinitely differentiable function. Consider the problem of finding a solution f to the differential equation

$$a_m \frac{d^mf}{dx^m} + a_{m-1} \frac{d^{m-1}}{dx^{m-1}} + \cdots + a_1 f = g.$$

We may rewrite this equation without the variable x, in the form

$$a_m D^m f + a_{m-1} D^{m-1} f + \cdots + a_1 f = g.$$

Each derivative D^k is a linear map from V to itself. Let

$$L = a_m D^m + a_{m-1} D^{m-1} + \cdots + a_1 I.$$

Then L is a sum of linear maps, and is itself a linear map. Thus the differential equation may be rewritten in the form

$$L(f) = g.$$

This is now in a similar notation to that used for solving linear equations. Furthermore, this equation is in "non-homogeneous" form. The associated homogeneous equation is the equation

$$L(f) = 0,$$

where the right-hand side is the zero function. Let W be the kernel of L. Then W is the set (space) of solutions of the homogeneous equation

$$a_m D^m f + \cdots + a_1 f = 0.$$

If there exists one solution f_0 for the non-homogeneous equation $L(f) = g$, then all solutions are obtained by the translation

$$f_0 + W = \text{set of all functions } f_0 + f \text{ with } f \text{ in } W.$$

See Exercise 5.

In several previous exercises we looked at the image of lines, planes, parallelograms under a linear map. For example, if we consider the plane spanned by two linearly independent vectors v_1, v_2 in V, and

$$L: V \to W$$

is a linear map, then the image of that plane will be a plane provided $L(v_1)$, $L(v_2)$ are also linearly independent. We can give a criterion for this in terms of the kernel, and the criterion is valid quite generally as follows.

Theorem 3.1. *Let $F: V \to W$ be a linear map whose kernel is $\{O\}$. If $v_1,\dots,v_n$ are linearly independent elements of V, then $F(v_1),\dots,F(v_n)$ are linearly independent elements of W.*

Proof. Let $x_1,\dots,x_n$ be numbers such that

$$x_1 F(v_1) + \cdots + x_n F(v_n) = O.$$

By linearity, we get

$$F(x_1 v_1 + \cdots + x_n v_n) = O.$$

Hence $x_1 v_1 + \cdots + x_n v_n = O$. Since $v_1,\dots,v_n$ are linearly independent it follows that $x_i = 0$ for $i = 1,\dots,n$. This proves our theorem.

We often abbreviate kernel and image by writing Ker and Im respectively. The next theorem relates the dimensions of the kernel and image of a linear map, with the dimension of the space on which the map is defined.

Theorem 3.2 *Let V be a vector space. Let $L: V \to W$ be a linear map of V into another space W. Let n be the dimension of V, q the dimension of the kernel of L, and s the dimension of the image of L. Then $n = q + s$. In other words,*

$$\dim V = \dim \text{Ker } L + \dim \text{Im } L.$$

Proof. If the image of L consists of O only, then our assertion is trivial. We may therefore assume that $s > 0$. Let $\{w_1,\dots,w_s\}$ be a basis of the image of L. Let $v_1,\dots,v_s$ be elements of V such that $L(v_i) = w_i$ for $i = 1,\dots,s$. If the kernel is not $\{O\}$, let $\{u_1,\dots,u_q\}$ be a basis of the kernel. If the kernel is $\{O\}$, it is understood that all reference to $\{u_1,\dots,u_q\}$ is to be omitted in what follows. We contend that

$$\{v_1,\dots,v_s,\ u_1,\dots,u_q\}$$

is a basis of V. This will suffice to prove our assertion. Let v be any element of V. Then there exist numbers $x_1, \ldots, x_s$ such that

$$L(v) = x_1 w_1 + \cdots + x_s w_s,$$

because $\{w_1, \ldots, w_s\}$ is a basis of the image of L. By linearity,

$$L(v) = L(x_1 v_1 + \cdots + x_s v_s),$$

and again by linearity, subtracting the right-hand side from the left-hand side, it follows that

$$L(v - x_1 v_1 - \cdots - x_s v_s) = O.$$

Hence $v - x_1 v_1 - \cdots - x_s v_s$ lies in the kernel of L, and there exist numbers $y_1, \ldots, y_q$ such that

$$v - x_1 v_1 - \cdots - x_s v_s = y_1 u_1 + \cdots + y_q u_q.$$

Hence

$$v = x_1 v_1 + \cdots + x_s v_s + y_1 u_1 + \cdots + y_q u_q$$

is a linear combination of $v_1, \ldots, v_s$, $u_1, \ldots, u_q$. This proves that these $s + q$ elements of V generate V.

We now show that they are linearly independent, and hence that they constitute a basis. Suppose that there exists a linear relation:

$$x_1 v_1 + \cdots + x_s v_s + y_1 u_1 + \cdots + y_q u_q = O.$$

Applying L to this relation, and using the fact that $L(u_j) = O$ for $j = 1, \ldots, q$, we obtain

$$x_1 L(v_1) + \cdots + x_s L(v_s) = O.$$

But $L(v_1), \ldots, L(v_s)$ are none other than $w_1, \ldots, w_s$, which have been assumed linearly independent. Hence $x_i = 0$ for $i = 1, \ldots, s$. Hence

$$y_1 u_1 + \cdots + y_q u_q = 0.$$

But $u_1, \ldots, u_q$ constitute a basis of the kernel of L, and in particular, are linearly independent. Hence all $y_j = 0$ for $j = 1, \ldots, q$. This concludes the proof of our assertion.

Example 1 (continued). The linear map $L: \mathbf{R}^3 \to \mathbf{R}$ of Example 1 is given by the formula

$$L(x, y, z) = 3x - 2y + z.$$

Its kernel consists of all solutions of the equation

$$3x - y + z = 0.$$

Its image is a subspace of **R**, is not $\{O\}$, and hence consists of all of **R**. Thus its image has dimension 1. Hence its kernel has dimension 2.

Example 2 (continued). The image of the projection

$$P: \mathbf{R}^3 \to \mathbf{R}^2$$

in Example 2 is all of $\mathbf{R}^2$, and the kernel has dimension 1.

Exercises IV, §3

Let $L: V \to W$ be a linear map.

1. (a) If S is a one-dimensional subspace of V, show that the image $L(S)$ is either a point or a line.
   (b) If S is a two-dimensional subspace of V, show that the image $L(S)$ is either a plane, a line or a point.

2. (a) If S is an arbitrary line in V (cf. Chapter III, §2) show that the image of S is either a point or a line.
   (b) If S is an arbitrary plane in V, show that the image of S is either a plane, a line or a point.

3. (a) Let $F: V \to W$ be a linear map, whose kernel is $\{O\}$. Assume that V and W have both the same dimension n. Show that the image of F is all of W.
   (b) Let $F: V \to W$ be a linear map and assume that the image of F is all of W. Assume that V and W have the same dimension n. Show that the kernel of F is $\{O\}$.

4. Let $L: V \to W$ be a linear map. Assume dim $V >$ dim W. Show that the kernel of L is not O.

5. Let $L: V \to W$ be a linear map. Let w be an element of W. Let v_0 be an element of V such that $L(v_0) = w$. Show that any solution of the equation $L(X) = w$ is of type $v_0 + u$, where u is an element of the kernel of L.

6. Let V be the vector space of functions which have derivatives of all orders, and let $D: V \to V$ be the derivative. What is the kernel of D?

7. Let D^2 be the second derivative (i.e. the iteration of D taken twice). What is the kernel of D^2? In general, what is the kernel of D^n (n-th derivative)?

8. (a) Let V, D be as in Exercise 6. Let $L = D - I$, where I is the identity mapping of V. What is the kernel of L?
   (b) Same question of $L = D - aI$, where a is a number.

9. (a) What is the dimension of the subspace of $\mathbf{R}^n$ consisting of those vectors $A = (a_1,\ldots,a_n)$ such that $a_1 + \cdots + a_n = 0$?

   (b) What is the dimension of the subspace of the space of $n \times n$ matrices (a_{ij}) such that

$$a_{11} + \cdots + a_{nn} = \sum_{i=1}^{n} a_{ii} = 0?$$

10. An $n \times n$ matrix A is called **skew-symmetric** if ${}^tA = -A$. Show that any $n \times n$ matrix A can be written as a sum

$$A = B + C,$$

where B is symmetric and C is skew-symmetric. [*Hint*: Let $B = (A + {}^tA)/2$.] Show that if $A = B_1 + C_1$, where B_1 is symmetric and C_1 is skew-symmetric, then $B = B_1$ and $C = C_1$.

11. Let M be the space of all $n \times n$ matrices. Let

$$P: M \to M$$

be the map such that

$$P(A) = \frac{A + {}^tA}{2}.$$

   (a) Show that P is linear.
   (b) Show that the kernel of P consists of the space of skew-symmetric matrices.
   (c) Show that the image of P consists of all symmetric matrices. [*Watch out.* You have to prove two things: For any matrix A, $P(A)$ is symmetric. Conversely, given a symmetric matrix B, there exists a matrix A such that $B = P(A)$. What is the simplest possibility for such A?]
   (d) You should have determined the dimension of the space of symmetric matrices previously, and found $n(n + 1)/2$. What then is the dimension of the space of skew-symmetric matrices?
   (e) Exhibit a basis for the space of skew-symmetric matrices.

12. Let M be the space of all $n \times n$ matrices. Let

$$Q: M \to M$$

be the map such that

$$Q(A) = \frac{A - {}^tA}{2}.$$

   (a) Show that Q is linear.
   (b) Describe the kernel of Q, and determine its dimension.
   (c) What is the image of Q?

13. A function (real valued, of a real variable) is called **even** if $f(-x) = f(x)$. It is called **odd** if $f(-x) = -f(x)$.
    (a) Verify that $\sin x$ is an odd function, and $\cos x$ is an even function.
    (b) Let V be the vector space of all functions. Define the map

$$P: V \to V$$

by $(Pf)(x) = (f(x) + f(-x))/2$. Show that P is a linear map.
    (c) What is the kernel of P?
    (d) What is the image of P? Prove your assertions.
14. Let again V be the vector space of all functions. Define the map

$$Q: V \to V$$

by $(Qf)(x) = f(x) - f(-x))/2$.
    (a) Show that Q is a linear map.
    (b) What is the kernel of Q?
    (c) What is the image of Q? Prove your assertion.

Remark. Exercises 11, 12, 13, 14 have certain formal elements in common. These common features will be discussed later. See Exercises 4 through 7 of Chapter V, §1.

15. **The product space.** Let U, W be vector spaces. We let the **direct product**, simply called the **product**, $U \times W$ be the set of all pairs (u, w) with $u \in U$ and $w \in W$. This should not be confused with the product of numbers, the scalar product, the cross product of vectors which is sometimes used in physics to denote a different type of operation. It is an unfortunate historical fact that the word product is used in two different contexts, and you should get accustomed to this. For instance, we can view $\mathbf{R}^4$ as a product,

$$\mathbf{R}^4 = \mathbf{R}^3 \times \mathbf{R}^1 = \mathbf{R}^3 \times \mathbf{R}$$

by viewing a 4-tuple (x_1, x_2, x_3, x_4) as putting side by side the triple (x_1, x_2, x_3) and the single number x_4. Similarly,

$$\mathbf{R}^4 = \mathbf{R}^2 \times \mathbf{R}^2,$$

by viewing (x_1, x_2, x_3, x_4) as putting side by side (x_1, x_2) and (x_3, x_4). If (u_1, w_1) and (u_2, w_2) are elements of $U \times W$, so

$$u_1, u_2 \in U \quad \text{and} \quad w_1, w_2 \in W,$$

we define their **sum** componentwise, that is we define

$$(u_1, w_1) + (u_2, w_2) = (u_1 + u_2, w_1 + w_2).$$

If c is a number, define $c(u, w) = (cu, cw)$.

(a) Show that $U \times W$ is a vector space with these definitions. What is the zero element?

(b) Show that $\dim(U \times W) = \dim U + \dim W$. In fact, let $\{u_i\}$ $(i = 1, \dots, n)$ be a basis of U and $\{w_j\}$ $(j = 1, \dots, m)$ be a basis of W. Show that the elements $\{(u_i, 0)\}$ and $\{(0, w_j)\}$ form a basis of $U \times W$.

(c) Let U be a subspace of a vector space V. Show that the subset of $V \times V$ consisting of all elements (u, u) with $u \in U$ is a subspace of $V \times V$.

(d) Let $\{u_i\}$ be a basis of U. Show that the set of elements (u_i, u_i) is a basis of the subspace in (c). Hence the dimension of this subspace is the same as the dimension of U.

16. (To be done after you have done Exercise 15.) Let U, W be subspaces of a vector space V. Show by the indicated method that

$$\boxed{\dim U + \dim W = \dim(U + W) + \dim(U \cap W).}$$

(a) Show that the map
$$L: U \times W \to V$$
given by
$$L(u, w) = u - w$$

is a linear map.

(b) Show that image of L is $U + W$.

(c) Show that the kernel of L is the subspace of $U \times W$ consisting of all elements (u, u) where u is in $U \cap W$. What is a basis for this subspace? What is its dimension?

(d) Apply the dimension formula in the text to conclude the proof.

IV, §4. The Rank and Linear Equations Again

Let A be an $m \times n$ matrix,

$$A = \begin{pmatrix} a_{11} & \cdots & a_{1n} \\ \vdots & & \vdots \\ a_{m1} & \cdots & a_{mn} \end{pmatrix}.$$

Let $L_A: \mathbf{R}^n \to \mathbf{R}^m$ be the linear map which has been defined previously, namely
$$L_A(X) = AX.$$

As we have mentioned, the kernel of L_A is the space of solutions of the system of linear equations written briefly as

$$AX = O.$$

Let us now analyze its image.

Let $E^1, \ldots, E^n$ be the standard unit vectors of $\mathbf{R}^n$, written as column vectors, so

$$E^1 = \begin{pmatrix} 1 \\ 0 \\ \vdots \\ 0 \end{pmatrix}, \quad \ldots, \quad E^n = \begin{pmatrix} 0 \\ 0 \\ \vdots \\ 1 \end{pmatrix}.$$

Then ordinary matrix multiplication shows that

$$AE^j = A^j$$

is the j-th column of A. Consequently for any vector

$$X = x_1 E^1 + \cdots + x_n E^n,$$

we find that

$$AX = L_A(X) = x_1 A^1 + \cdots + x_n A^n.$$

Thus we see:

Theorem 4.1. *The image of L_A is the subspace generated by the columns of A.*

In Chapter III, we gave a name to the dimension of that space, namely the **column rank**, which we have already seen is equal to the row rank, and is simply called the **rank** of A. Now we can interpret this rank also in the following way:

> The rank of A is the dimension of the image of L_A.

Theorem 4.2. *Let r be the rank of A. Then the dimension of the space of solutions of $AX = O$ is equal to $n - r$.*

Proof. By Theorem 3.2 we have

$$\dim \operatorname{Im} L_A + \dim \operatorname{Ker} L_A = n.$$

But $\dim \operatorname{Im} L_A = r$ and $\operatorname{Ker} L_A$ is the space of solutions of the homogeneous linear equations, so our assertion is now clear.

Example 1. Find the dimension of the space of solutions of the system of equations

$$2x - y + z + 2w = 0,$$

$$x + y - 2z - w = 0.$$

Here the matrix A is

$$\begin{pmatrix} 2 & -1 & 1 & 2 \\ 1 & 1 & -2 & -1 \end{pmatrix}.$$

It has rank 2 because the two vectors

$$\begin{pmatrix} 2 \\ 1 \end{pmatrix} \quad \text{and} \quad \begin{pmatrix} -1 \\ 1 \end{pmatrix}$$

are easily seen to be linearly independent. [Either use row and column operations, or do this by linear equations.] Hence the dimension of the space of solutions is $4 - 2 = 2$.

We recall that the system of linear equations could also be written in the form

$$X \cdot A_i = 0 \qquad \text{for} \quad i = 1, \dots, m,$$

where A_i are the rows of the matrix A. This means that X is perpendicular to each row of A. Then X is also perpendicular to the row space of A, i.e. to the space generated by the rows. It is now convenient to introduce some terminology.

Let U be a subspace of $\mathbf{R}^n$. We let

$U^\perp$ = set of all elements X in $\mathbf{R}^n$ such that $X \cdot Y = 0$ for all Y in U.

We call $U^\perp$ the **orthogonal complement** of U. It is the set of vectors which are perpendicular to all elements of U, or as we shall also say, perpendicular to U itself. Then it is easily verified that $U^\perp$ is a subspace (Exercise 8).

Let U be the subspace generated by the row vectors of the matrix $A = (a_{ij})$. Then its orthogonal complement $U^\perp$ is precisely the set of solutions of the homogeneous equations

$$X \cdot A_i = 0 \qquad \text{for all } i.$$

In other words, we have

$$\boxed{\text{(row space of } A)^\perp = \text{Ker } L_A = \text{space of solutions of } AX = O.}$$

Theorem 4.3. *Let U be a subspace of $\mathbf{R}^n$. Then*

$$\dim U + \dim U^\perp = n.$$

Proof. Let $r = \dim U$. If $r = 0$, then the assertion is obvious. If $r \neq 0$ then U has a basis, and in particular is generated by a finite number of vectors $A_1, \ldots, A_r$, which may be viewed as the rows of a matrix. Then the dimension formula is a special case of Theorem 4.2.

In 3-dimensional space, for instance, Theorem 4.3 proves the fact that the orthogonal complement of a line is a plane, and vice versa, as shown on the figure.

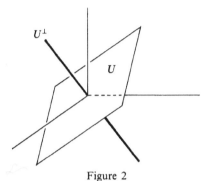

Figure 2

In 4-space, the orthogonal complement of a subspace of dimension 1 has dimension 3. The orthogonal complement of a subspace of dimension 2 has also dimension 2.

Let us now discuss briefly non-homogeneous equations, i.e. a system of the form

$$AX = B,$$

where B is a given vector (m-tuple). Such a system may not have a solution, in other words, the equations may be what is called "inconsistent".

Example 2. Consider the system

$$3x - y + z = 1,$$
$$2x + y - z = 2,$$
$$x - 2y + 2z = 5.$$

It turns out that the third row of the matrix of coefficients

$$A = \begin{pmatrix} 3 & -1 & 1 \\ 2 & 1 & -1 \\ 1 & -2 & 2 \end{pmatrix}$$

is obtained by subtracting the second row from the first. Hence it follows at once that the rank of the matrix is 2. On the other hand, $5 \neq 1 - 2$ so there cannot be a solution to the above system of equations.

Theorem 4.4. *Consider a non-homogeneous system of linear equations*

$$AX = B.$$

Suppose that there exists at least one solution X_0. Then the set of solutions is precisely

$$X_0 + \text{Ker } L_A.$$

In other words, all the solutions are of the form

$$X_0 + Y, \quad \text{where } Y \text{ is a solution of } \quad AY = 0.$$

Proof. Let $Y \in \text{Ker } L_A$. This means $AY = 0$. Then

$$A(X_0 + Y) = AX_0 + AY = B + 0 = B.$$

so $X_0 + \text{Ker } L_A$ is contained in the set of solutions. Conversely, let X be any solution of $AX = B$. Then

$$A(X - X_0) = AX - AX_0 = B - B = 0.$$

Hence $X = X_0 + (X - X_0)$, where $X - X_0 = Y$ and $AY = 0$. This proves the theorem.

When there exists one solution at least to the system $AX = B$, then $\dim \text{Ker } L_A$ is called the **dimension of the set of solutions**. It is the dimension of the homogeneous system.

Example 3. Find the dimension of the set of solutions of the following system of equations, and determine this set in $\mathbf{R}^3$.

$$2x + y + z = 1,$$
$$y - z = 0.$$

We see by inspection that there is at least one solution, namely $x = \frac{1}{2}$, $y = z = 0$. The rank of the matrix

$$\begin{pmatrix} 2 & 1 & 1 \\ 0 & 1 & -1 \end{pmatrix}$$

is 2. Hence the dimension of the set of solutions is 1. The vector space of solutions of the homogeneous system has dimension 1, and one solution is easily found to be

$$y = z = 1, \qquad x = -1.$$

Hence the set of solutions of the inhomogeneous system is the set of all vectors

$$(\tfrac{1}{2}, 0, 0) + t(-1, 1, 1),$$

where t ranges over all real numbers. We see that our set of solutions is a straight line.

Exercises IV, §4

1. Let A be a non-zero vector in $\mathbf{R}^n$. What is the dimension of the space of solutions of the equation $A \cdot X = 0$?

2. What is the dimension of the subspace of $\mathbf{R}^6$ perpendicular to the two vectors $(1, 1, -2, 3, 4, 5)$ and $(0, 0, 1, 1, 0, 7)$?

3. Let A be a non-zero vector in n-space. Let P be a point in n-space. What is the dimension of the set of solutions of the equation

$$X \cdot A = P \cdot A?$$

4. What is the dimension of the space of solutions of the following systems of linear equations? In each case, find a basis for the space of solutions.
   (a) $2x + y - z = 0$ (b) $x - y + z = 0$
       $2x + y + z = 0$
   (c) $4x + 7y - \pi z = 0$ (d) $x + y + z = 0$
       $2x - y + z = 0$ $x - y \quad = 0$
                                    $y + z = 0$

5. What is the dimension of the space of solutions of the following systems of linear equations?
   (a) $2x - 3y + z = 0$ (b) $2x + 7y = 0$
        $x + y - z = 0$ $x - 2y + z = 0$
   (c) $2x - 3y + z = 0$ (d) $x + y + z = 0$
        $x + y - z = 0$ $2x + 2y + 2z = 0$
        $3x + 4y = 0$
        $5x + y + z = 0$

6. Let $L : V \to W$ be a linear map. Using a theorem from the text, prove that

$$\dim \operatorname{Im} L \le \dim V.$$

7. Let A, B be two matrices which can be multiplied, i.e. such that AB exists. Prove that

$$\operatorname{rank} AB \le \operatorname{rank} A \quad \text{and} \quad \operatorname{rank} AB \le \operatorname{rank} B.$$

8. Let U be a subspace of $\mathbf{R}^n$. Prove that $U^\perp$ is also a subspace.

IV, §5. The Matrix Associated with a Linear Map

To every matrix A we have associated a linear map L_A. Conversely, given a linear map

$$L: \mathbf{R}^n \to \mathbf{R}^m,$$

we shall now prove that there is some associated matrix A such that $L = L_A$.

Let $E^1, \ldots, E^n$ be the unit column vectors of $\mathbf{R}^n$. For each $j = 1, \ldots, n$ let $L(E^j) = A^j$, where A^j is a column vector in $\mathbf{R}^m$. Thus

$$L(E^1) = \begin{pmatrix} a_{11} \\ \vdots \\ a_{m1} \end{pmatrix} = A^1, \ldots, \qquad L(E^n) = \begin{pmatrix} a_{1n} \\ \vdots \\ a_{mn} \end{pmatrix} = A^n.$$

Then for every element X in $\mathbf{R}^n$ we can write

$$X = x_1 E^1 + \cdots + x_n E^n = \begin{pmatrix} x_1 \\ \vdots \\ x_n \end{pmatrix}$$

and therefore

$$
\begin{aligned}
L(X) &= x_1 L(E^1) + \cdots + x_n L(E^n) \\
&= x_1 A^1 + \cdots + x_n A^n \\
&= AX
\end{aligned}
$$

where A is the matrix whose column vectors are $A^1, \ldots, A^n$. Hence $L = L_A$, which proves the theorem.

Remark. When dealing with $\mathbf{R}^n$ and $\mathbf{R}^m$, we are able to write column vectors, so the matrix A was easily derived above. Later in this section we deal with more general vector spaces, in terms of bases, and we shall write coordinate vectors horizontally. This will give rise to a transpose, due to the horizontal notation.

The matrix A above will be called the matrix **associated** with the linear map L.

As we had seen in studying the column space of A, we can express the columns of A in terms of the images of the unit vectors:

(∗)
$$L(E^1) = \begin{pmatrix} a_{11} \\ \vdots \\ a_{m1} \end{pmatrix}, \quad \ldots, \quad L(E^n) = \begin{pmatrix} a_{1n} \\ \vdots \\ a_{mn} \end{pmatrix}.$$

Example 1. Let $F: \mathbf{R}^3 \to \mathbf{R}^2$ be the projection, in other words the mapping such that $F^t(x_1, x_2, x_3) = {}^t(x_1, x_2)$. Then the matrix associated with F is

$$\begin{pmatrix} 1 & 0 & 0 \\ 0 & 1 & 0 \end{pmatrix}.$$

Example 2. Let $I: \mathbf{R}^n \to \mathbf{R}^n$ be the identity. Then the matrix associated with I is the matrix

$$\begin{pmatrix} 1 & 0 & 0 & \cdots & 0 \\ 0 & 1 & 0 & \cdots & 0 \\ \vdots & \vdots & \vdots & & \vdots \\ 0 & 0 & 0 & \cdots & 1 \end{pmatrix},$$

having components equal to 1 on the diagonal, and 0 otherwise.

Example 3. Let $L: \mathbf{R}^4 \to \mathbf{R}^2$ be the linear map such that

$$L(E^1) = \begin{pmatrix} 2 \\ 1 \end{pmatrix}, \quad L(E^2) = \begin{pmatrix} 3 \\ -1 \end{pmatrix}, \quad L(E^3) = \begin{pmatrix} -5 \\ 4 \end{pmatrix}, \quad L(E^4) = \begin{pmatrix} 1 \\ 7 \end{pmatrix}.$$

According to the relations (∗), we see that the matrix associated with L is the matrix

$$\begin{pmatrix} 2 & 3 & -5 & 1 \\ 1 & -1 & 4 & 7 \end{pmatrix}.$$

Remark. If instead of column vectors we used row vectors, then to find the associated matrix would give rise to a *transpose*.

Let V be an n-dimensional vector space. If we pick some basis $\{v_1, \ldots, v_n\}$ of V, then every element of V can be written in terms of coordinates

$$v = x_1 v_1 + \cdots + x_n v_n.$$

Thus to each element v of V we can associate the **coordinate vector**

$$X = \begin{pmatrix} x_1 \\ \vdots \\ x_n \end{pmatrix}.$$

If

$$w = y_1 v_1 + \cdots + y_n v_n$$

so Y is the coordinate vector of w, then

$$v + w = (x_1 + y_1)v_1 + \cdots + (x_n + y_n)v_n,$$

so $X + Y$ is the coordinate vector of $v + w$. Let c be a number. Then

$$cX = cx_1 v_1 + \cdots + cx_n v_n,$$

so cX is the coordinate vector of cv. Thus after choosing a basis, we can identify V with $\mathbf{R}^n$ via the coordinate vectors.

Let $L: V \to V$ be a linear map. Then after choosing a basis which gives us an identification of V with $\mathbf{R}^n$, we can then represent L by a matrix. Different choices of bases will give rise to different associated matrices. Some choices of bases will often give rise to especially simple matrices.

Example. Suppose that there exists a basis $\{v_1, \ldots, v_n\}$ and numbers $c_1, \ldots, c_n$ such that

$$Lv_i = c_i v_i \qquad \text{for} \qquad i = 1, \ldots, n.$$

Then with respect to this basis, the matrix of L is the diagonal matrix

$$\begin{pmatrix} c_1 & 0 & \cdots & 0 \\ 0 & c_2 & \cdots & 0 \\ \vdots & \vdots & & \vdots \\ 0 & 0 & \cdots & c_n \end{pmatrix}.$$

If we picked another basis, the matrix of L might not be so simple.

The general principle for finding the associated matrix of a linear map with respect to a basis can be found as follows.

Let $\{v_1, \ldots, v_n\}$ be the given basis of V. Then there exist numbers c_{ij} such that

$$
\begin{aligned}
Lv_1 &= c_{11}v_1 + \cdots + c_{1n}v_n \\
&\vdots \qquad \vdots \qquad\qquad \vdots \\
Lv_n &= c_{n1}v_1 + \cdots + c_{nn}v_n.
\end{aligned}
$$

What is the effect of L on the coordinate vector X of an element $v \in V$? Such an element is of the form

$$v = x_1 v_1 + \cdots + x_n v_n.$$

Then

$$Lv = \sum_{i=1}^{n} x_i L(v_i)$$

$$= \sum_{i=1}^{n} x_i \sum_{j=1}^{n} c_{ij} v_j$$

$$= \sum_{i=1}^{n} \sum_{j=1}^{n} x_i c_{ij} v_j$$

$$= \sum_{j=1}^{n} \left(\sum_{i=1}^{n} x_i c_{ij} \right) v_j.$$

Hence we find:

If $C = (c_{ij})$ is the matrix such that $L(v_i) = \sum_{j=1}^{n} c_{ij} v_j$, and X is the coordinate vector of v, then the coordinate vector of Lv is ${}^t CX$. In other words, on coordinate vectors, L is represented by the matrix ${}^t C$ (transpose of C).

We note the *transpose* of C rather than C itself. This is because when writing Lv_i as linear combination of $v_1, \ldots, v_n$ we have written it horizontally, whereas before we wrote it vertically in terms of the vertical unit vectors $E^1, \ldots, E^n$. We call ${}^t C$ the **matrix associated** with L with respect to the given basis.

Example. Let $L: V \to V$ be a linear map. Let $\{v_1, v_2, v_3\}$ be a basis of V such that

$$L(v_1) = 2v_1 - v_2,$$
$$L(v_2) = v_1 + v_2 - 4v_3,$$
$$L(v_3) = 5v_1 + 4v_2 + 2v_3.$$

Then the matrix associated with L on coordinate vectors is the matrix

$$\begin{pmatrix} 2 & 1 & 5 \\ -1 & 1 & 4 \\ 0 & -4 & 2 \end{pmatrix}.$$

It is the *transpose* of the matrix

$$\begin{pmatrix} 2 & -1 & 0 \\ 1 & 1 & -4 \\ 5 & 4 & 2 \end{pmatrix}.$$

Appendix: Change of Bases

You may also ask how the matrix representing a linear map changes when we change a basis of V. We can easily find the answer as follows. First we discuss how coordinates change.

Let $\{v_1, \ldots, v_n\}$ be one basis, and $\{w_1, \ldots, w_n\}$ another basis of V.

Let X denote the coordinates of a vector with respect to $\{v_1, \ldots, v_n\}$ and let Y denote the coordinates of the same vector with respect to $\{w_1, \ldots, w_n\}$.

How do X and Y differ? We shall now give the answer. Let v be the element of V having coordinates X, viewed as a column vector. Thus

$$v = x_1 v_1 + \cdots + x_n v_n.$$

This looks like a dot product, and it will be convenient to use the notation

$$v = {}^t\!X \begin{pmatrix} v_1 \\ \vdots \\ v_n \end{pmatrix} = (x_1, \ldots, x_n) \begin{pmatrix} v_1 \\ \vdots \\ v_n \end{pmatrix},$$

where ${}^t\!X$ is now a row n-tuple. Similarly,

$$v = {}^t\!Y \begin{pmatrix} w_1 \\ \vdots \\ w_n \end{pmatrix} = (y_1, \ldots, y_n) \begin{pmatrix} w_1 \\ \vdots \\ w_n \end{pmatrix}.$$

We can express each w_i as a linear combination of the basis elements $v_1, \ldots, v_n$ so there exists a matrix $B = (b_{ij})$ such that for each i,

$$\boxed{w_i = \sum_{j=1}^{n} b_{ij} v_j.}$$

But we can write these relations more efficiently in matrix form

$$\begin{pmatrix} w_1 \\ \vdots \\ w_n \end{pmatrix} = \begin{pmatrix} b_{11} & \cdots & b_{1n} \\ \vdots & & \vdots \\ b_{n1} & \cdots & b_{nn} \end{pmatrix} \begin{pmatrix} v_1 \\ \vdots \\ v_n \end{pmatrix} = B \begin{pmatrix} v_1 \\ \vdots \\ v_n \end{pmatrix}.$$

Therefore the relation for v in terms of bases elements can be written in the form

$$v = {}^tY \begin{pmatrix} w_1 \\ \vdots \\ w_n \end{pmatrix} = {}^tYB \begin{pmatrix} v_1 \\ \vdots \\ v_n \end{pmatrix} \qquad \text{and also} \qquad v = {}^tX \begin{pmatrix} v_1 \\ \vdots \\ v_n \end{pmatrix}.$$

Therefore we must have ${}^tYB = {}^tX$. Taking the transpose gives us the desired relation

$$\boxed{X = {}^tBY.}$$

Again notice the transpose. The change of coordinates from one basis to another are given in terms of the transpose of the matrix expressing each w_i as a linear combination of $v_1, \ldots, v_n$.

Remark. *The matrix B is invertible.*

Proof. There are several ways of seeing this. For instance, by the same arguments we have given, going from one basis to another, there is a matrix C such that

$$Y = {}^tCX.$$

This is true for all coordinate n-tuples X and Y. Thus we obtain

$$X = {}^tB{}^tCX = {}^t(CB)X$$

for all n-tuples X. Hence $CB = I$ is the identity matrix. Similarly $BC = I$, so B is invertible.

Now let $L: V \to V$ be a linear map. Let M be the matrix representing L with respect to the basis $\{v_1, \ldots, v_n\}$ and let M' be the matrix representing L with respect to the basis $\{w_1, \ldots, w_n\}$. By definition,

the coordinates of $L(v)$ are $\begin{cases} MX \text{ with respect to } \{v_1, \ldots, v_n\}, \\ M'Y \text{ with respect to } \{w_1, \ldots, w_n\}. \end{cases}$

By what we have just seen, we must have

$$MX = {}^tBM'Y.$$

Substitute $X = {}^t\!BY$ and multiply both sides on the left by ${}^t\!B^{-1}$. Then we find

$$ {}^t\!B^{-1}M{}^t\!BY = M'Y. $$

This is true for all Y. If we let $N = {}^t\!B$ then we obtain the matrix M' in terms of M, namely

$$ \boxed{M' = N^{-1}MN} \qquad \text{where} \quad N = {}^t\!B. $$

Thus the matrix representing the linear map changes by a similarity transformation. We may also say that M, M' are similar. In general, two matrices M, M' are called **similar** if there exists an invertible matrix N such that $M' = N^{-1}MN$.

In practice, one should not pick bases too quickly. For many problems one should select a basis for which the matrix representing the linear map is simplest, and work with that basis.

Example. Suppose that with respect to some basis the matrix M representing L is diagonal, say

$$ M = \begin{pmatrix} 2 & 0 \\ 0 & -1 \end{pmatrix}. $$

Then the matrix representing L with respect to another basis will be of the form

$$ N^{-1}MN, $$

which may look like a horrible mess. Changing N arbitrarily corresponds to picking an arbitrary basis (of course, N must be invertible). When we study eigenvectors later, we shall find conditions under which a matrix representing a linear map is diagonal with respect to a suitable choice of basis.

Exercises IV, §5

1. Find the matrix associated with the following linear maps.
    (a) $F: \mathbf{R}^4 \to \mathbf{R}^2$ given by $F^t(x_1, x_2, x_3, x_4) = {}^t(x_1, x_2)$ (the projection).
    (b) The projection from $\mathbf{R}^4$ to $\mathbf{R}^3$.
    (c) $F: \mathbf{R}^2 \to \mathbf{R}^2$ given by $F^t(x, y) = {}^t(3x, 3y)$.
    (d) $F: \mathbf{R}^n \to \mathbf{R}^n$ given by $F(X) = 7X$.
    (e) $F: \mathbf{R}^n \to \mathbf{R}^n$ given by $F(X) = -X$.
    (f) $F: \mathbf{R}^4 \to \mathbf{R}^4$ given by $F^t(x_1, x_2, x_3, x_4) = {}^t(x_1, x_2, 0, 0)$.

2. Let c be a number, and let $L: \mathbf{R}^n \to \mathbf{R}^n$ be the linear map such that $L(X) = cX$. What is the matrix associated with this linear map?

3. Let $F: \mathbf{R}^3 \to \mathbf{R}^2$ be the indicated linear map. What is the associated matrix of F?

(a) $F(E^1) = \begin{pmatrix} 1 \\ -3 \end{pmatrix}$, $F(E^2) = \begin{pmatrix} -4 \\ 2 \end{pmatrix}$, $F(E^3) = \begin{pmatrix} 3 \\ 1 \end{pmatrix}$.

(b) $F \begin{pmatrix} x_1 \\ x_2 \\ x_3 \end{pmatrix} = \begin{pmatrix} 3x_1 - 2x_2 + x_3 \\ 4x_1 - x_2 + 5x_3 \end{pmatrix}$.

4. Let V be a 3-dimensional space with basis $\{v_1, v_2, v_3\}$. Let $F: V \to V$ be the linear map as indicated. Find the matrix of F with respect to the given basis.

(a) $F(v_1) = 3v_2 - v_3$,
   $F(v_2) = v_1 - 2v_2 + v_3$,
   $F(v_3) = -2v_1 + 4v_2 + 5v_3$.

(b) $F(v_1) = 3v_1$, $F(v_2) = -7v_2$, $F(v_3) = 5v_3$.

(c) $F(v_1) = -2v_1 + 7v_3$,
   $F(v_2) = -v_3$,
   $F(v_3) = v_1$.

5. In the text, we gave a description of a matrix associated with a linear map of a vector space into itself, with respect to a basis. More generally, let V, W be two vector spaces. Let $\{v_1, \ldots, v_n\}$ be a basis of V, and $\{w_1, \ldots, w_m\}$ a basis of W. Let $L: V \to W$ be a linear map. Describe how you would associate a matrix with this linear map, giving the effect on coordinate vectors.

6. Let $L: V \to V$ be a linear map. Let $v \in V$. We say that v is an eigenvector for L if there exists a number c such that $L(v) = cv$. Suppose that V has a basis $\{v_1, \ldots, v_n\}$ consisting of eigenvectors, with $L(v_i) = c_i v_i$ for $i = 1, \ldots, n$. What is the matrix representing L with respect to this basis?

7. Let V be the vector space generated by the two functions $f_1(t) = \cos t$ and $f_2(t) = \sin t$. Let D be the derivative. What is the matrix of D with respect to the basis $\{f_1, f_2\}$?

8. Let V be the vector space generated by the three functions $f_1(t) = 1$, $f_2(t) = t$, $f_3(t) = t^2$. Let $D: V \to V$ be the derivative. What is the matrix of D with respect to the basis $\{f_1, f_2, f_3\}$?

CHAPTER V

Composition and Inverse Mappings

V, §1. Composition of Linear Maps

Let U, V, W be sets. Let

$$F: U \to V \quad \text{and} \quad G: V \to W$$

be mappings. Then we can form the composite mapping from U into W, denoted by $G \circ F$. It is by definition the mapping such that

$$(G \circ F)(u) = G(F(u)) \quad \text{for all } u \text{ in } U.$$

Example 1. Let A be an $m \times n$ matrix, and let B be a $q \times m$ matrix. Then we may form the product BA. Let

$$L_A: \mathbf{R}^n \to \mathbf{R}^m \quad \text{be the linear map such that} \quad L_A(X) = AX$$

and let

$$L_B: \mathbf{R}^m \to \mathbf{R}^q \quad \text{be the linear map such that} \quad L_B(Y) = BY.$$

Then we may form the composite linear map $L_B \circ L_A$ such that

$$(L_B \circ L_A)(X) = L_B(L_A(X)) = L_B(AX) = BAX.$$

Thus we have

$$\boxed{L_B \circ L_A = L_{BA}.}$$

We see that composition of linear maps corresponds to multiplication of matrices.

Example 2. Let A be an $m \times n$ matrix, and let

$$L_A: \mathbf{R}^n \to \mathbf{R}^m$$

be the usual linear map such that $L_A(X) = AX$. Let C be a vector in $\mathbf{R}^m$ and let

$$T_C: \mathbf{R}^m \to \mathbf{R}^m$$

be the translation by C, that is $T_C(Y) = Y + C$. Then the composite mapping $T_C \circ L_A$ is obtained by first applying L_A to a vector X, and then translating by C. Thus

$$T_C \circ L_A(X) = T_C(L_A(X)) = T_C(AX) = AX + C.$$

Example 3. Let V be a vector space, and let w be an element of V. Let

$$T_w: V \to V$$

be the translation by w, that is the map such that $T_w(v) = v + w$. Then we have

$$T_{w_1}(T_{w_2}(v)) = T_{w_1}(v + w_2) = v + w_2 + w_1.$$

Thus

$$T_{w_1} \circ T_{w_2} = T_{w_1 + w_2}.$$

We can express this by saying that the composite of two translations is again a translation. Of course, the translation T_w is not a linear map if $w \neq O$ because

$$T_w(O) = O + w = w \neq O,$$

and we know that a linear map has to send O on O.

Example 4. Rotations. Let θ be a number, and let $A(\theta)$ be the matrix

$$A(\theta) = \begin{pmatrix} \cos\theta & -\sin\theta \\ \sin\theta & \cos\theta \end{pmatrix}.$$

Then $A(\theta)$ represents a rotation which we may denote by R_θ. The composite rotation $R_{\theta_1} \circ R_{\theta_2}$ is obtained from the multiplication of matrices, and for any vector X in $\mathbf{R}^2$ we have

$$R_{\theta_1} \circ R_{\theta_2}(X) = A(\theta_1)A(\theta_2)X.$$

This composite rotation is just rotation by the sum of the angles, namely $\theta_1 + \theta_2$. This corresponds to the formula $A(\theta_1)A(\theta_2) = A(\theta_1 + \theta_2)$.

The following statement is an important property of mappings.

Let U, V, W, S be sets. Let

$$F: U \to V, \qquad G: V \to W, \qquad and \qquad H: W \to S$$

be mappings. Then

$$H \circ (G \circ F) = (H \circ G) \circ F.$$

Proof. Here again, the proof is very simple. By definition, we have, for any element u of U:

$$(H \circ (G \circ F))(u) = H((G \circ F)(u)) = H(G(F(u))).$$

On the other hand,

$$((H \circ G) \circ F)(u) = (H \circ G)(F(u)) = H(G(F(u))).$$

By definition, this means that $(H \circ G) \circ F = H \circ (G \circ F)$.

Theorem 1.1. *Let U, V, W be vector spaces. Let*

$$F: U \to V \qquad and \qquad G: V \to W$$

be linear maps. Then the composite map $G \circ F$ is also a linear map.

Proof. This is very easy to prove. Let u, v be elements of U. Since F is linear, we have $F(u + v) = F(u) + F(v)$. Hence

$$(G \circ F)(u + v) = G(F(u + v)) = G(F(u) + F(v)).$$

Since G is linear, we obtain

$$G(F(u) + F(v)) = G(F(u)) + G(F(v)).$$

Hence

$$(G \circ F)(u + v) = (G \circ F)(u) + (G \circ F)(v).$$

Next, let c be a number. Then

$$
\begin{aligned}
(G \circ F)(cu) &= G(F(cu)) \\
&= G(cF(u)) \quad \text{(because F is linear)} \\
&= cG(F(u)) \quad \text{(because G is linear).}
\end{aligned}
$$

This proves that $G \circ F$ is a linear mapping.

The next theorem states that some of the rules of arithmetic concerning the product and sum of numbers also apply to the composition and sum of linear mappings.

Theorem 1.2. *Let U, V, W be vector spaces. Let*

$$F: U \to V$$

be a linear mapping, and let G, H be two linear mappings of V into W. Then

$$(G + H) \circ F = G \circ F + H \circ F.$$

If c is a number, then

$$(cG) \circ F = c(G \circ F).$$

If $T: U \to V$ is a linear mapping from U into V, then

$$G \circ (F + T) = G \circ F + G \circ T.$$

The proofs are all simple. We shall just prove the first assertion and leave the others as exercises.

Let u be an element of U. We have:

$$
\begin{aligned}
((G + H) \circ F)(u) = (G + H)(F(u)) &= G(F(u)) + H(F(u)) \\
&= (G \circ F)(u) + (H \circ F)(u).
\end{aligned}
$$

By definition, it follows that $(G + H) \circ F = G \circ F + H \circ F$.

As with matrices, we see that composition and addition of linear maps behaves like multiplication and addition of numbers. However, the same warning as with matrices applies here. First, we may not have commutativity, and second we do not have "division", except as discussed in the next section for inverses, when they exist.

Example 5. Let

$$F: \mathbf{R}^3 \to \mathbf{R}^3$$

be the linear map given by

$$F(x, y, z) = (x, y, 0)$$

and let G be the linear mapping given by

$$G(x, y, z) = (x, z, 0)$$

Then $(G \circ F)(x, y, z) = (x, 0, 0)$, but $(F \circ G)(x, y, z) = (x, z, 0)$.

On the other hand, let

$$L: V \to V$$

be a linear map of a vector space into itself. We may iterate L several times, so as usual we let

$$L^2 = L \circ L, \qquad L^3 = L \circ L \circ L, \qquad \text{and so forth.}$$

We also let

$$L^0 = I = \text{identity mapping.}$$

Thus L^k is the iteration of L with itself k times. For such powers of L, we do have commutativity, namely

$$L^{r+s} = L^r \circ L^s = L^s \circ L^r.$$

Exercises V, §1

1. Let A, B be two $m \times n$ matrices. Assume that

$$AX = BX$$

for all n-tuples X. Show that $A = B$.

2. Let F, L be linear maps of V into itself. Assume that F, L commute, that is $F \circ L = L \circ F$. Prove the usual rules:

$$(F + L)^2 = F^2 + 2F \circ L + L^2,$$
$$(F - L)^2 = F^2 - 2F \circ L + L^2,$$
$$(F + L) \circ (F - L) = F^2 - L^2.$$

3. Prove the usual rule for a linear map $F: V \to V$:

$$(I - F) \circ (I + F + \cdots + F^r) = I - F^{r+1}.$$

4. Let V be a vector space and $T: V \to V$ a linear map such that $T^2 = I$. Define $P = \frac{1}{2}(I + T)$ and $Q = \frac{1}{2}(I - T)$.
   (a) Show that $P^2 = P$, and $Q^2 = Q$.
   (b) Show that $P + Q = I$.
   (c) Show that $\operatorname{Ker} P = \operatorname{Im} Q$ and $\operatorname{Im} P = \operatorname{Ker} Q$.

5. Let $P: V \to V$ be a linear map such that $P^2 = P$. Define $Q = I - P$.
   (a) Show that $Q^2 = Q$.
   (b) Show that $\operatorname{Im} P = \operatorname{Ker} Q$ and $\operatorname{Ker} P = \operatorname{Im} Q$.
   A linear map P such that $P^2 = P$ is called a **projection**. It generalizes the notion of projection in the usual sense.

6. Let $P: V \to V$ be a **projection**, that is a linear map such that $P^2 = P$.
   (a) Show that $V = \operatorname{Ker} P + \operatorname{Im} P$.
   (b) Show that the intersection of $\operatorname{Im} P$ and $\operatorname{Ker} P$ is $\{O\}$. In other words, if $v \in \operatorname{Im} P$ and $v \in \operatorname{Ker} P$, then $v = O$.

Let V be a vector space, and let U, W be subspaces. One says that V is a **direct sum** of U and W if the following conditions are satisfied:

$$V = U + W \quad \text{and} \quad U \cap W = \{O\}.$$

In Exercise 6, you have proved that V is the direct sum of $\operatorname{Ker} P$ and $\operatorname{Im} P$.

7. Let V be the direct sum of subspaces U and W. Let $v \in V$ and suppose we have expressed v as a sum $v = u + w$ with $u \in U$ and $w \in W$. Show that u, w are uniquely determined by v. That is, if $v = u_1 + w_1$ with $u_1 \in U$ and $w_1 \in W$, then $u = u_1$ and $w = w_1$.

8. Let U, W be two vector spaces, and let $V = U \times W$ be the set of all pairs (u, w) with $u \in U$ and $w \in W$. Then V is a vector space, as described in Exercise 15 of Chapter IV, §3. Let
$$P: V \to V$$
be the map such that $P(u, w) = (u, 0)$. Show that P is a projection.

If you identify U with the set of all elements $(u, 0)$ with $u \in U$, and identify W with the set of all elements $(0, w)$ with $w \in W$, then V is the direct sum of U and W.

For example, let $n = r + s$ where r, s are positive integers. Then

$$\mathbf{R}^n = \mathbf{R}^r \times \mathbf{R}^s.$$

Note that $\mathbf{R}^n$ is the set of all n-tuples of real numbers, which can be viewed as

$$(x_1, \ldots, x_r, y_1, \ldots, y_s), \quad \text{with} \quad x_i, y_j \in \mathbf{R}.$$

Thus $\mathbf{R}^n$ may be viewed as the direct sum of $\mathbf{R}^r$ and $\mathbf{R}^s$. The projection of $\mathbf{R}^n$ on the first r components is given by the map P such that

$$P(x_1, \ldots, x_r, y_1, \ldots, y_s) = (x_1, \ldots, x_r, 0, \ldots, 0).$$

This map is a linear map, and $P^2 = P$. Sometimes one also calls the map such that

$$P(x_1,\ldots,x_n) = (x_1,\ldots,x_r)$$

a projection, but it maps $\mathbf{R}^n$ into $\mathbf{R}^r$, so we cannot again apply P to $(x_1,\ldots,x_r)$ since P is defined on all of $\mathbf{R}^n$.

Note that we could also take the projection on the second set of coordinates, that is, we let

$$Q(x_1,\ldots,x_r, y_1,\ldots,y_s) = (0,\ldots,0, y_1,\ldots,y_s).$$

This is called the projection on the second factor of $\mathbf{R}^n$, viewed as $\mathbf{R}^r \times \mathbf{R}^s$.

9. Let A, B be two matrices so that the product AB is defined. Prove that

$$\text{rank } AB \leq \text{rank } A \qquad \text{and} \qquad \text{rank } AB \leq \text{rank } B.$$

[*Hint*: Consider the linear maps L_{AB}, L_A, and L_B.]

In fact, define the **rank of a linear map** L to be dim Im L. If $L: V \to W$ and $F: W \to U$ are linear maps of finite dimensional spaces, show that

$$\text{rank } F \circ L \leq \text{rank } F \qquad \text{and} \qquad \text{rank } F \circ L \leq \text{rank } L.$$

10. Let A be an $n \times n$ matrix, and let C be a vector in $\mathbf{R}^n$. Let T_C be the translation by C as in Example 2. Write out in full the formulas for

$$L_A \circ T_C(X) \qquad \text{and} \qquad T_C \circ L_A(X),$$

where X is in $\mathbf{R}^n$. Give an example of A and C such that

$$L_A \circ T_C \neq T_C \circ L_A.$$

V, §2. Inverses

Let

$$F: V \to W$$

be a mapping (which in the applications is linear). We say that F has an **inverse** if there exists a mapping

$$G: W \to V$$

such that

$$G \circ F = I_V \qquad \text{and} \qquad F \circ G = I_W.$$

By this we mean that the composite maps $G \circ F$ and $F \circ G$ are the identity mappings of V and W, respectively. If F has an inverse, we also say that F is **invertible**.

Example 1. The inverse mapping of the translation T_u is the translation T_{-u}, because

$$T_{-u} \circ T_u(v) = T_{-u}(v + u) = v + u - u = v.$$

Thus

$$T_{-u} \circ T_u = I.$$

Similarly, $T_u \circ T_{-u} = I.$

Example 2. Let A be a square $n \times n$ matrix, and let

$$L_A : \mathbf{R}^n \to \mathbf{R}^n$$

be the usual linear map such that $L_A(X) = AX$. Assume that A has an inverse matrix A^{-1}, so that $AA^{-1} = A^{-1}A = I$. Then the formula

$$L_A \circ L_B = L_{AB}$$

of the preceding sections shows that

$$L_A \circ L_{A^{-1}} = L_I = I.$$

Hence L_A has an inverse mapping, which is precisely multiplication by A^{-1}.

Theorem 2.1. *Let $F : U \to V$ be a linear map, and assume that this map has an inverse mapping $G : V \to U$. Then G is a linear map.*

Proof. Let $v_1, v_2 \in V$. We must first show that

$$G(v_1 + v_2) = G(v_1) + G(v_2).$$

Let $u_1 = G(v_1)$ and $u_2 = G(v_2)$. By definition, this means that

$$F(u_1) = v_1 \qquad \text{and} \qquad F(u_2) = v_2.$$

Since F is linear, we find that

$$F(u_1 + u_2) = F(u_1) + F(u_2) = v_1 + v_2.$$

By definition of the inverse map, this means that $G(v_1 + v_2) = u_1 + u_2$, thus proving what we wanted. We leave the proof that $G(cv) = cG(v)$ as an exercise (Exercise 2).

Example 3. Let $L: V \to V$ be a linear map such that $L^2 = O$. Then $I + L$ is invertible, because

$$(I + L)(I - L) = I^2 - L^2 = I,$$

and similarly on the other side, $(I - L)(I + L) = I$. Thus we have

$$(I + L)^{-1} = I - L.$$

We shall now express inverse mappings in somewhat different terminology.

Let

$$F: V \to W$$

be a mapping. We say that F is **injective** (in older terminology, **one-to-one**) if given elements v_1, v_2 in V such that $v_1 \neq v_2$ then $F(v_1) \neq F(v_2)$.

We are mostly interested in this notion for linear mappings.

Example 4. Suppose that F is a linear map whose kernel is not $\{O\}$. Then there is an element $v \neq O$ in the kernel, and we have

$$F(O) = F(v) = O.$$

Hence F is not injective. We shall now prove the converse.

Theorem 2.2. *A linear map $F: V \to W$ is injective if and only if its kernel is $\{O\}$.*

Proof. We have already proved one implication. Conversely, assume that the kernel is $\{O\}$. We must prove that F is injective. Let $v_1 \neq v_2$ be distinct elements of V. We must show that $F(v_1) \neq F(v_2)$. But

$$F(v_1) - F(v_2) = F(v_1 - v_2) \quad \text{because } F \text{ is linear.}$$

Since the kernel of F is $\{O\}$, and $v_1 - v_2 \neq O$, it follows that $F(v_1 - v_2) \neq O$. Hence $F(v_1) - F(v_2) \neq O$, so $F(v_1) \neq F(v_2)$. This proves the theorem.

Let $F: V \to W$ be a mapping. If the image of F is all of W then we say that F is **surjective**. The two notions for a mapping to be injective or surjective combine to give a basic criterion for F to have an inverse.

Theorem 2.3. *A mapping $F: V \to W$ has an inverse if and only if it is both injective and surjective.*

Proof. Suppose F is both injective and surjective. Given an element w in W, there exists an element v in V such that $F(v) = w$ (because F is surjective). There is only one such element v (because F is injective). Thus we may define

$$G(w) = \text{unique element } v \text{ such that } F(v) = w.$$

By the way we have defined G, it is then clear that

$$G(F(v)) = v \quad \text{and} \quad F(G(w)) = w.$$

Thus G is the inverse mapping of F.

Conversely, suppose F has an inverse mapping G. Let v_1, v_2 be elements of V such that $F(v_1) = F(v_2)$. Applying G yields

$$v_1 = G \circ F(v_1) = G \circ F(v_2) = v_2,$$

so F is injective. Secondly, let w be an element of W. The equation

$$w = F(G(w))$$

shows that $w = F(v)$ for some v, namely $v = G(w)$, so F is surjective. This proves the theorem.

In the case of linear maps, we have certain tests for injectivity, or surjectivity, which allow us to verify fewer conditions when we wish to prove that a linear map is invertible.

Theorem 2.4. *Let* $F: V \to W$ *be a linear map. Assume that*

$$\dim V = \dim W.$$

(i)   *If* $\text{Ker } F = \{O\}$ *then* F *is invertible.*
(ii) *If* F *is surjective, then* F *is invertible.*

Proof. Suppose first that $\text{Ker } F = \{O\}$. Then F is injective by Theorem 2.2. But

$$\dim V = \dim \text{Ker } F + \dim \text{Im } F,$$

so $\dim V = \dim \text{Im } F$, and the image of F is a subspace of W having the same dimension as W. Hence $\text{Im } F = W$ by Theorem 5.6 of Chapter III. Hence F is surjective. This proves (i), by using Theorem 2.3.

The proof of (ii) will be left as an exercise.

Example 5. Let $F: \mathbf{R}^2 \to \mathbf{R}^2$ be the linear map such that

$$F(x, y) = (3x - y, 4x + 2y).$$

We wish to show that F has an inverse. First note that the kernel of F is $\{O\}$, because if

$$3x - y = 0,$$

$$4x + 2y = 0,$$

then we can solve for x, y in the usual way: Multiply the first equation by 2 and add it to the second. We find $10x = 0$, whence $x = 0$, and then $y = 0$ because $y = 3x$. Hence F is injective, because its kernel is $\{O\}$.

Hence F is invertible by Theorem 2.4(i).

A linear map $F: U \to V$ which has an inverse $G: V \to U$ (we also say **invertible**) is called an **isomorphism**.

Example 6. Let V be a vector space of dimension n. Let

$$\{v_1, \ldots, v_n\}$$

be a basis for V. Let

$$L: \mathbf{R}^n \to V$$

be the map such that

$$L(x_1, \ldots, x_n) = x_1 v_1 + \cdots + x_n v_n.$$

Then L is an isomorphism.

Proof. The kernel of L is $\{O\}$, because if

$$x_1 v_1 + \cdots + x_n v_n = O,$$

then all $x_i = 0$ (since $v_1, \ldots, v_n$ are linearly independent). The image of L is all of V, because $v_1, \ldots, v_n$ generate V. By Theorem 2.4, it follows that L is an isomorphism.

Theorem 2.5. *A square matrix A is invertible if and only if its columns $A^1, \ldots, A^n$ are linearly independent.*

Proof. By Theorem 2.4, the linear map L_A is invertible if and only if $\operatorname{Ker} L_A = \{O\}$. But $\operatorname{Ker} L_A$ consists of those n-tuples X such that

$$x_1 A^1 + \cdots + x_n A^n = O,$$

in other words, those X giving a relation of linear dependence among the columns. Hence the theorem follows.

Exercises V, §2

1. Let R_θ be rotation counterclockwise by an angle θ. How would you express the inverse R_θ^{-1} as R_φ for some φ in a simple way? If

$$A(\theta) = \begin{pmatrix} \cos\theta & -\sin\theta \\ \sin\theta & \cos\theta \end{pmatrix}$$

is the matrix associated with R_θ, what is the matrix associated with R_θ^{-1}?

2. (a) Finish the proof of Theorem 2.1.
   (b) Give the proof of Theorem 2.4(ii).

3. Let F, G be invertible linear maps of a vector space V onto itself. Show that

$$(F \circ G)^{-1} = G^{-1} \circ F^{-1}.$$

4. Let $L: \mathbf{R}^2 \to \mathbf{R}^2$ be the linear map defined by

$$L(x, y) = (x + y, x - y).$$

Show that L is invertible.

5. Let $L: \mathbf{R}^2 \to \mathbf{R}^2$ be the linear map defined by

$$L(x, y) = (2x + y, 3x - 5y).$$

Show that L is invertible.

6. Let $L: \mathbf{R}^3 \to \mathbf{R}^3$ be the linear maps as indicated. Show that L is invertible in each case.
   (a) $L(x, y, z) = (x - y, x + z, x + y + 3z)$
   (b) $L(x, y, z) = (2x - y + z, x + y, 3x + y + z)$.

7. Let $L: V \to V$ be a linear mapping such that $L^2 = O$. Show that $I - L$ is invertible. (I is the identity mapping on V.)

8. Let $L: V \to V$ be a linear map such that $L^2 + 2L + I = O$. Show that L is invertible.

9. Let $L: V \to V$ be a linear map such that $L^3 = O$. Show that $I - L$ is invertible.

10. Let $L: V \to V$ be a linear map such that $L^r = O$. Show that $I - L$ is invertible.

11. Let V be a two-dimensional vector space, and let $L: V \to V$ be a linear map such that $L^2 = O$, but $L \neq O$. Let v be an element of V such that $L(v) \neq O$. Let $w = L(v)$. Prove that $\{v, w\}$ is a basis of V.

12. Let V be the set of all infinite sequences of real numbers

$$(x_1, x_2, x_3, \ldots).$$

This could be called infinite dimensional space. Addition and multiplication by numbers are defined componentwise, so V is a vector space. Define the map $F: V \to V$ by

$$F(x_1, x_2, x_3, \ldots) = (0, x_1, x_2, x_3, \ldots).$$

For obvious reasons, F is called the **shift operator**, and F is linear.
(a) Is F injective? What is the kernel of F?
(b) Is F surjective?
(c) Show that there is a linear map $G: V \to V$ such that $G \circ F = I$.
(d) Does the map G of (c) have the property that $F \circ G = I$?

13. Let V be a vector space, and let U, W be two subspaces. Assume that V is the direct sum of U and W, that is

$$V = U + W \qquad \text{and} \qquad U \cap W = \{O\}.$$

Let $L: U \times W \to V$ be the mapping such that

$$L(u, w) = u + w.$$

Show that L is a bijective linear map. (So prove that L is linear, L is injective, L is surjective.)

CHAPTER VI

Scalar Products and Orthogonality

VI, §1. Scalar Products

Let V be a vector space. A **scalar product** on V is an association which to any pair of elements (v, w) of V associates a number, denoted by $\langle v, w \rangle$, satisfying the following properties:

SP 1. *We have* $\langle v, w \rangle = \langle w, v \rangle$ *for all* v, w *in* V.

SP 2. *If* u, v, w *are elements of* V, *then*

$$\langle u, v + w \rangle = \langle u, v \rangle + \langle u, w \rangle.$$

SP 3. *If* x *is a number, then*

$$\langle xu, v \rangle = x\langle u, v \rangle = \langle u, xv \rangle.$$

We shall also assume that the scalar product satisfies the condition:

SP 4. *For all* v *in* V *we have* $\langle v, v \rangle \geq 0$, *and* $\langle v, v \rangle > 0$ *if* $v \neq O$.

A scalar product satisfying this condition is called **positive definite**.

For the rest of this section we assume that V is a vector space with a positive definite scalar product.

Example 1. Let $V = \mathbf{R}^n$, and define

$$\langle X, Y \rangle = X \cdot Y$$

for elements X, Y of $\mathbf{R}^n$. Then this is a positive definite scalar product.

Example 2. Let V be the space of continuous real-valued functions on the interval $[-\pi, \pi]$. If f, g are in V, we define

$$\langle f, g \rangle = \int_{-\pi}^{\pi} f(t)g(t) \, dt.$$

Simple properties of the integral show that this is a scalar product, which is in fact positive definite.

In calculus, we study the second example, which gives rise to the theory of Fourier series. Here we discuss only general properties of scalar products and applications to euclidean spaces. The notation $\langle \ , \ \rangle$ is used because in dealing with vector spaces of functions, a dot $f \cdot g$ may be confused with the ordinary product of functions.

As in the case of the dot product, we define elements v, w of V to be **orthogonal**, or **perpendicular**, and write $v \perp w$, if $\langle v, w \rangle = 0$. If S is a subset of V, we denote by $S^\perp$ the set of all elements w in V which are perpendicular to all elements of S, i.e. such that $\langle w, v \rangle = 0$ for all v in S. Then using **SP 1**, **SP 2**, and **SP 3**, one verifies at once that $S^\perp$ is a subspace of V, called the **orthogonal space** of S. If w is perpendicular to S, we also write $w \perp S$. Let U be the subspace of V generated by the elements of S. If w is perpendicular to S, and if v_1, v_2 are in S, then

$$\langle w, v_1 + v_2 \rangle = \langle w, v_1 \rangle + \langle w, v_2 \rangle = 0.$$

If c is a number, then

$$\langle w, cv_1 \rangle = c\langle w, v_1 \rangle = 0.$$

Hence w is perpendicular to linear combinations of elements of S, and hence w is perpendicular to U.

Example 3. Let (a_{ij}) be an $m \times n$ matrix, and let $A_1, \ldots, A_m$ be its row vectors. Let $X = (x_1, \ldots, x_n)$ as usual. The system of homogeneous linear equations

$$a_{11}x_1 + \cdots + a_{1n}x_n = 0$$
(**)
$$\vdots \qquad \vdots \qquad \vdots$$
$$a_{m1}x_1 + \cdots + a_{mn}x_n = 0$$

can also be written in abbreviated form using the dot product, as

$$A_1 \cdot X = 0, \quad \ldots, \quad A_m \cdot X = 0.$$

The set of solutions X of this homogeneous system is therefore the set of all vectors perpendicular to $A_1, \ldots, A_m$. It is therefore the subspace of $\mathbf{R}^n$ which is the orthogonal subspace to the space generated by $A_1, \ldots, A_m$. If U is the space of solutions, and if W denotes the space generated by $A_1, \ldots, A_m$, we have

$$U = W^\perp.$$

We call dim U the **dimension of the space of solutions of the system of linear equations**.

As in Chapter I, we define the **length**, or **norm** of an element $v \in V$ by

$$\|v\| = \sqrt{\langle v, v \rangle}.$$

If c is any number, then we immediately get

$$\|cv\| = |c| \, \|v\|,$$

because

$$\|cv\| = \sqrt{\langle cv, cv \rangle} = \sqrt{c^2 \langle v, v \rangle} = |c| \, \|v\|.$$

Thus we see the same type of arguments as in Chapter I apply here. In fact, any argument given in Chapter I which does not use coordinates applies to our more general situation. We shall see further examples as we go along.

As before, we say that an element $v \in V$ is a **unit vector** if $\|v\| = 1$. If $v \in V$ and $v \neq O$, then $v/\|v\|$ is a unit vector.

The following two identities follow directly from the definition of the length.

The Pythagoras theorem. *If v, w are perpendicular, then*

$$\|v + w\|^2 = \|v\|^2 + \|w\|^2.$$

The parallelogram law. *For any v, w we have*

$$\|v + w\|^2 + \|v - w\|^2 = 2\|v\|^2 + 2\|w\|^2.$$

The proofs are trivial. We give the first, and leave the second as an exercise. For the first, we have

$$\|v + w\|^2 = \langle v + w, v + w \rangle = \langle v, v \rangle + 2\langle v, w \rangle + \langle w, w \rangle$$
$$= \|v\|^2 + \|w\|^2.$$

Let w be an element of V such that $\|w\| \neq 0$. For any v there exists a unique number c such that $v - cw$ is perpendicular to w. Indeed, for $v - cw$ to be perpendicular to w we must have

$$\langle v - cw, w \rangle = 0,$$

whence $\langle v, w \rangle - \langle cw, w \rangle = 0$ and $\langle v, w \rangle = c\langle w, w \rangle$. Thus

$$c = \frac{\langle v, w \rangle}{\langle w, w \rangle}.$$

Conversely, letting c have this value shows that $v - cw$ is perpendicular to w. We call c the **component of v along** w.

In particular, if w is a unit vector, then the component of v along w is simply

$$c = \langle v, w \rangle.$$

Example 4. Let $V = \mathbf{R}^n$ with the usual scalar product, i.e. the dot product. If E_i is the i-th unit vector, and $X = (x_1, \ldots, x_n)$ then the component of X along E_i is simply

$$X \cdot E_i = x_i,$$

that is, the i-th component of X.

Example 5. Let V be the space of continuous functions on $[-\pi, \pi]$. Let f be the function given by $f(x) = \sin kx$, where k is some integer > 0. Then

$$\|f\| = \sqrt{\langle f, f \rangle} = \left(\int_{-\pi}^{\pi} \sin^2 kx \, dx \right)^{1/2}$$
$$= \sqrt{\pi}.$$

If g is any continuous function on $[-\pi, \pi]$, then the component of g along f is also called the **Fourier coefficient of g with respect to** f, and is equal to

$$\frac{\langle g, f \rangle}{\langle f, f \rangle} = \frac{1}{\pi} \int_{-\pi}^{\pi} g(x) \sin kx \, dx.$$

As with the case of n-space, we define the **projection of v along w** to be the vector cw, because of our usual picture:

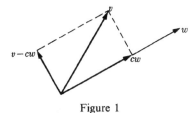

Figure 1

Exactly the same arguments which we gave in Chapter I can now be used to get the **Schwarz inequality,** namely:

Theorem 1.1 *For all v, $w \in V$ we have*

$$|\langle v, w \rangle| \leq \|v\| \, \|w\|.$$

Proof. If $w = 0$, then both sides are equal to 0 and our inequality is obvious. Next, assume that $w \neq 0$. Let c be the component of v along w. We write

$$v = v - cw + cw.$$

Then $v - cw$ is perpendicular to cw, so by Pythagoras,

$$\|v\|^2 = \|v - cw\|^2 + \|cw\|^2$$
$$= \|v - cw\|^2 + |c|^2 \, \|w\|^2.$$

Therefore $|c|^2 \, \|w\|^2 \leq \|v\|^2$ and taking square roots yields

$$|c| \, \|w\| \leq \|v\|.$$

But $c = \langle v, w \rangle / \|w\|^2$. Then one factor $\|w\|$ cancels, and cross multiplying by $\|w\|$ yields

$$|\langle v, w \rangle| \leq \|v\| \, \|w\|,$$

thereby proving the theorem.

Theorem 1.2 *If v, $w \in V$, then*

$$\|v + w\| \leq \|v\| + \|w\|.$$

Proof. Exactly the same as that of the analogous theorem in Chapter I, §4.

Let $v_1,\ldots,v_n$ be non-zero elements of V which are mutually perpendicular, that is $\langle v_i, v_j \rangle = 0$ if $i \neq j$. Let c_i be the component of v along v_i. Then

$$v - c_1 v_1 - \cdots - c_n v_n$$

is perpendicular to $v_1,\ldots,v_n$. To see this, all we have to do is to take the product with v_j for any j. All the terms involving $\langle v_i, v_j \rangle$ will give 0 if $i \neq j$, and we shall have two remaining terms

$$\langle v, v_j \rangle - c_j \langle v_j, v_j \rangle$$

which cancel. Thus subtracting linear combinations as above orthogonalizes v with respect to $v_1,\ldots,v_n$. The next theorem shows that $c_1 v_1 + \cdots + c_n v_n$ gives the closest approximation to v as a linear combination of $v_1,\ldots,v_n$.

Theorem 1.3 Let $v_1,\ldots,v_n$ be vectors which are mutually perpendicular, and such that $\|v_i\| \neq 0$ for all i. Let v be an element of V, and let c_i be the component of v along v_i. Let $a_1,\ldots,a_n$ be numbers. Then

$$\left\| v - \sum_{k=1}^{n} c_k v_k \right\| \leqq \left\| v - \sum_{k=1}^{n} a_k v_k \right\|.$$

Proof. We know that

$$v - \sum_{k=1}^{n} c_k v_k$$

is perpendicular to each v_i, $i = 1,\ldots,n$. Hence it is perpendicular to any linear combination of $v_1,\ldots,v_n$. Now we have:

$$\|v - \sum a_k v_k\|^2 = \|v - \sum c_k v_k + \sum (c_k - a_k) v_k\|^2$$
$$= \|v - \sum c_k v_k\|^2 + \| \sum (c_k - a_k) v_k\|^2$$

by the Pythagoras theorem. This proves that

$$\|v - \sum c_k v_k\|^2 \leqq \|v - \sum a_k v_k\|^2,$$

and thus our theorem is proved.

Example 6. Consider the vector space V of all continuous functions on the interval $[0, 2\pi]$. Let

$$g_k(x) = \cos kx, \qquad \text{for} \quad k = 0, 1, 2,\ldots.$$

We use the scalar product

$$\langle f, g \rangle = \int_0^{2\pi} f(x)g(x)\ dx.$$

Then it is easily verified that

$$\|g_0\| = \sqrt{2\pi} \quad \text{and} \quad \|g_k\| = \sqrt{\pi} \quad \text{for} \quad k > 0.$$

The Fourier coefficient of f with respect to g_k is

$$c_k = \frac{1}{\pi} \int_0^{2\pi} f(x)\ \cos\ kx,\ dx, \quad \text{for} \quad k > 0.$$

If we take $v_k = g_k$ for $k = 1,\ldots,n$ then Theorem 1.3 tells us that the linear combination

$$c_0 + c_1 \cos x + c_2 \cos 2x + \cdots + c_n \cos nx$$

gives the best approximation to the function f among all possible linear combinations

$$a_0 + a_1 \cos x + \cdots + a_n \cos nx$$

with arbitrary real numbers $a_0, a_1,\ldots,a_n$. Such a sum is called a **partial sum** of the Fourier series.

Similarly, we could take linear combinations of the functions $\sin kx$. This leads into the theory of Fourier series. We do not go into this deeper here. We merely wanted to point out the analogy, and the usefulness of the geometric language and formalism in dealing with these objects.

The next theorem is known as the **Bessel inequality**.

Theorem 1.4. *If $v_1,\ldots,v_n$ are mutually perpendicular unit vectors, and if c_i is the Fourier coefficient of v with respect to v_i, then*

$$\sum_{i=1}^n c_i^2 \leq \|v\|^2.$$

Proof. We have

$$0 \leq \langle v - \sum c_i v_i, v - \sum c_i v_i \rangle$$
$$= \langle v, v \rangle - \sum 2c_i \langle v, v_i \rangle + \sum c_i^2$$
$$= \langle v, v \rangle - \sum c_i^2.$$

From this our inequality follows.

Exercises VI, §1

1. Let V be a vector space with a positive definite scalar product. Let $v_1, \ldots, v_r$ be non-zero elements of V which are mutually perpendicular, meaning that $\langle v_i, v_j \rangle = 0$ if $i \neq j$. Show that $v_1, \ldots, v_r$ are linearly independent.

The following exercise gives an important example of a scalar product.

2. Let A be a symmetric $n \times n$ matrix. Given two column vectors X, $Y \in \mathbf{R}^n$, define

$$\langle X, Y \rangle = {}^t X A Y.$$

(a) Show that this symbol satisfies the first three properties of a scalar product.

(b) Give an example of a 1×1 matrix and a non-zero 2×2 matrix such that the fourth property is not satisfied. If this fourth property is satisfied, that is ${}^t X A X > 0$ for all $X \neq O$, then the matrix A is called **positive definite**.

(c) Give an example of a 2×2 matrix which is symmetric and positive definite.

(d) Let $a > 0$, and let

$$A = \begin{pmatrix} a & b \\ b & d \end{pmatrix}.$$

Prove that A is positive definite if and only if $ad - b^2 > 0$. [*Hint:* Let $X = {}^t(x, y)$ and complete the square in the expression ${}^t X A X$.]

(e) If $a < 0$ show that A is not positive definite.

3. Determine whether the following matrices are positive definite.

(a) $\begin{pmatrix} 3 & -1 \\ -1 & 2 \end{pmatrix}$ (b) $\begin{pmatrix} -2 & 1 \\ 1 & 5 \end{pmatrix}$

(c) $\begin{pmatrix} 4 & 1 \\ 1 & 2 \end{pmatrix}$ (d) $\begin{pmatrix} 4 & 4 \\ 4 & 1 \end{pmatrix}$

(e) $\begin{pmatrix} 4 & 1 \\ 1 & 10 \end{pmatrix}$ (f) $\begin{pmatrix} 4 & -1 \\ -1 & 10 \end{pmatrix}$

The trace of a matrix

4. Let A be an $n \times n$ matrix. Define the **trace** of A to be the sum of the diagonal elements. Thus if $A = (a_{ij})$, then

$$\operatorname{tr}(A) = \sum_{i=1}^{n} a_{ii}.$$

For instance, if

$$A = \begin{pmatrix} 1 & 2 \\ 3 & 4 \end{pmatrix},$$

then $\operatorname{tr}(A) = 1 + 4 = 5$. If

$$A = \begin{pmatrix} 1 & -1 & 5 \\ 2 & 1 & 3 \\ 1 & -4 & 7 \end{pmatrix},$$

then $\operatorname{tr}(A) = 9$. Compute the trace of the following matrices:

(a) $\begin{pmatrix} 1 & 7 & 3 \\ -1 & 5 & 2 \\ 2 & 3 & -4 \end{pmatrix}$ (b) $\begin{pmatrix} 3 & -2 & 4 \\ 1 & 4 & 1 \\ -7 & -3 & -3 \end{pmatrix}$ (c) $\begin{pmatrix} -2 & 1 & 1 \\ 3 & 4 & 4 \\ -5 & 2 & 6 \end{pmatrix}$.

5. (a) For any square matrix A show that $\operatorname{tr}(A) = \operatorname{tr}({}^tA)$.
   (b) Show that the trace is a linear map.

6. If A is a symmetric square matrix, show that $\operatorname{tr}(AA) \geq 0$, and $= 0$ if and only if $A = O$.

7. Let A, B be the indicated matrices. Show that
$$\operatorname{tr}(AB) = \operatorname{tr}(BA).$$

(a) $A = \begin{pmatrix} 1 & -1 & 1 \\ 2 & 4 & 1 \\ 3 & 0 & 1 \end{pmatrix}$ $B = \begin{pmatrix} 3 & 1 & 2 \\ 1 & 1 & 0 \\ -1 & 2 & 1 \end{pmatrix}$

(b) $A = \begin{pmatrix} 1 & 7 & 3 \\ -1 & 5 & 2 \\ 2 & 3 & -4 \end{pmatrix}$ $B = \begin{pmatrix} 3 & -2 & 4 \\ 1 & 4 & 1 \\ -7 & -3 & 2 \end{pmatrix}$.

8. (a) Prove in general that if A, B are square $n \times n$ matrices, then

$$\operatorname{tr}(AB) = \operatorname{tr}(BA).$$

(b) If C is an $n \times n$ matrix which has an inverse, then $\operatorname{tr}(C^{-1}AC) = \operatorname{tr}(A)$.

9. Let V be the vector space of symmetric $n \times n$ matrices. For $A, B \in V$ define the symbol

$$\langle A, B \rangle = \operatorname{tr}(AB),$$

where tr is the trace (sum of the diagonal elements). Show that the previous properties in particular imply that this defines a positive definite scalar product on V.

Exercises 10 through 13 deal with the scalar product in the context of calculus.

10. Let V be the space of continuous functions on $[0, 2\pi]$, and let the scalar product be given by the integral over this interval as in the text, that is

$$\langle f, g \rangle = \int_0^{2\pi} f(x)g(x)\ dx.$$

Let $g_n(x) = \cos nx$ for $n \geq 0$ and $h_m(x) = \sin mx$ for $m \geq 1$.

(a) Show that $\|g_0\| = \sqrt{2\pi}$, $\|g_n\| = \|h_n\| = \sqrt{\pi}$ for $n \geq 1$.

(b) Show that $g_n \perp g_m$ if $m \neq n$ and that $g_n \perp h_m$ for all m, n. *Hint:* Use formulas like

$$\sin A \cos B = \tfrac{1}{2}[\sin(A + B) + \sin(A - B)]$$

$$\cos A \cos B = \tfrac{1}{2}[\cos(A + B) + \cos(A - B)].$$

11. Let $f(x) = x$ on the interval $[0, 2\pi]$. Find $\langle f, g_n \rangle$ and $\langle f, h_n \rangle$ for the functions g_n, h_n of Exercise 10. Find the Fourier coefficients of f with respect to g_n and h_n.

12. Same question as in Exercise 11 if $f(x) = x^2$. (Exercises 10 through 13 give you a review of some elementary integrals from calculus.)

13. (a) Let $f(x) = x$ on the interval $[0, 2\pi]$. Find $\|f\|$.

(b) Let $f(x) = x^2$ on the same interval. Find $\|f\|$.

VI, §2. Orthogonal Bases

Let V be a vector space with a positive definite scalar product throughout this section. A basis $\{v_1, \ldots, v_n\}$ of V is said to be **orthogonal** if its elements are mutually perpendicular, i.e. if $\langle v_i, v_j \rangle = 0$ whenever $i \neq j$. If in addition each element of the basis has norm 1, then the basis is called **orthonormal**.

Example 1. The standard unit vectors

$$E_1, \ldots, E_n \quad \text{in} \quad \mathbf{R}^n$$

form an orthonormal basis of $\mathbf{R}^n$. Indeed, each has norm 1, and they are mutually orthogonal, that is

$$E_i \cdot E_j = 0 \quad \text{if} \quad i \neq j.$$

Of course there are many other orthonormal bases of $\mathbf{R}^n$.

This example is typical in the following sense.

Let $\{e_1, \ldots, e_n\}$ be an orthonormal basis of V. Any vector $v \in V$ can be written in terms of coordinates

$$v = x_1 e_1 + \cdots + x_n e_n \quad \text{with} \quad x_i \in \mathbf{R}.$$

Let w be another element of V, and write

$$w = y_1 e_1 + \cdots + y_n e_n \quad \text{with} \quad y_i \in \mathbf{R}.$$

Then

$$\langle v, w \rangle = \langle x_1 e_1 + \cdots + x_n e_n, y_1 e_1 + \cdots + y_n e_n \rangle$$

$$= \sum_{i,j=1}^{n} \langle x_i e_i, y_j e_j \rangle$$

$$= \sum_{i=1}^{n} x_i y_i$$

because $\langle e_i, e_j \rangle = 0$ for $i \neq j$. Hence if X is the coordinate n-tuple of v and Y the coordinate n-tuple of w, then

$$\langle v, w \rangle = X \cdot Y$$

so the scalar product is given precisely as the dot product of the coordinates. This is one of the uses of orthonormal bases: to identify the scalar product with the old-fashioned dot product.

Example 2. Consider $\mathbf{R}^2$. Let

$$A = (1, 1) \qquad \text{and} \qquad B = (1, -1).$$

Then $A \cdot B = 0$, so A is orthogonal to B, and A, B are linearly independent. Therefore they form a basis of $\mathbf{R}^2$, and in fact they form an orthogonal basis of $\mathbf{R}^2$. To get an orthonormal basis from them, we divide each by its norm, so an orthonormal basis is given by

$$\left(\frac{1}{\sqrt{2}}, \frac{1}{\sqrt{2}} \right) \qquad \text{and} \qquad \left(\frac{1}{\sqrt{2}}, \frac{-1}{\sqrt{2}} \right).$$

In general, suppose we have a subspace W of $\mathbf{R}^n$, and let $A_1, \ldots, A_r$ be any basis of W. We want to get an orthogonal basis of W. We follow a stepwise orthogonalization process. We start with $A_1 = B_1$. Then we take A_2 and subtract its projection on A_1 to get a vector B_2. Then we take A_3 and subtract its projections on B_1 and B_2 to get a vector B_3. Then we take A_4 and subtract its projections on B_1, B_2, B_3 to get a vector B_4. We continue in this way. This will eventually lead to an orthogonal basis of W.

We state this as a theorem and prove it in the context of vector spaces with a scalar product.

Theorem 2.1. Let V be a finite dimensional vector space, with a positive definite scalar product. Let W be a subspace of V, and let $\{w_1, \ldots, w_m\}$ be an orthogonal basis of W. If $W \neq V$, then there exist elements $w_{m+1}, \ldots, w_n$ of V such that $\{w_1, \ldots, w_n\}$ is an orthogonal basis of V.

Proof. The method of proof is as important as the theorem, and is called the **Gram–Schmidt orthogonalization process**. We know from Chapter III, §3 that we can find elements $v_{m+1}, \ldots, v_n$ of V such that

$$\{w_1, \ldots, w_m, \, v_{m+1}, \ldots, v_n\}$$

is a basis of V. Of course, it is not an orthogonal basis. Let W_{m+1} be the space generated by $w_1, \ldots, w_m, \, v_{m+1}$. We shall first obtain an orthogonal basis of W_{m+1}. The idea is to take v_{m+1} and subtract from it its projection along $w_1, \ldots, w_m$. Thus we let

$$c_1 = \frac{\langle v_{m+1}, w_1 \rangle}{\langle w_1, w_1 \rangle}, \quad \ldots, \quad c_m = \frac{\langle v_{m+1}, w_m \rangle}{\langle w_m, w_m \rangle}.$$

Let

$$w_{m+1} = v_{m+1} - c_1 w_1 - \cdots - c_m w_m.$$

Then w_{m+1} is perpendicular to $w_1, \ldots, w_m$. Furthermore, $w_{m+1} \neq O$ (otherwise v_{m+1} would be linearly dependent on $w_1, \ldots, w_m$), and v_{m+1} lies in the space generated by $w_1, \ldots, w_{m+1}$ because

$$v_{m+1} = w_{m+1} + c_1 w_1 + \cdots + c_m w_m.$$

Hence $\{w_1, \ldots, w_{m+1}\}$ is an orthogonal basis of W_{m+1}. We can now proceed by induction, showing that the space W_{m+s} generated by

$$w_1, \ldots, w_m, \, v_{m+1}, \ldots, v_{m+s}$$

has orthogonal basis

$$\{w_1, \ldots, w_{m+1}, \ldots, w_{m+s}\}$$

with $s = 1, \ldots, n - m$. This concludes the proof.

Corollary 2.2. *Let V be a finite dimensional vector space with a positive definite scalar product. Assume that $V \neq \{O\}$. Then V has an orthogonal basis.*

Proof. By hypothesis, there exists an element v_1 of V such that $v_1 \neq O$. We let W be the subspace generated by v_1, and apply the theorem to get the desired basis.

We summarize the procedure of Theorem 2.1 once more. Suppose we are given an arbitrary basis $\{v_1,\ldots,v_n\}$ of V. We wish to orthogonalize it. We proceed as follows. We let

$$v_1' = v_1,$$

$$v_2' = v_2 - \frac{\langle v_2, v_1' \rangle}{\langle v_1', v_1' \rangle}\, v_1',$$

$$v_3' = v_3 - \frac{\langle v_3, v_2' \rangle}{\langle v_2', v_2' \rangle}\, v_2' - \frac{\langle v_3, v_1' \rangle}{\langle v_1', v_1' \rangle}\, v_1',$$

$$\vdots \qquad\qquad \vdots$$

$$v_n' = v_n - \frac{\langle v_n, v_{n-1}' \rangle}{\langle v_{n-1}', v_{n-1}' \rangle}\, v_{n-1}' - \cdots - \frac{\langle v_n, v_1' \rangle}{\langle v_1', v_1' \rangle}\, v_1'.$$

Then $\{v_1',\ldots,v_n'\}$ is an orthogonal basis.

Given an orthogonal basis, we can always obtain an orthonormal basis by dividing each vector by its norm.

Example 3. Find an orthonormal basis for the vector space generated by the vectors $(1, 1, 0, 1)$, $(1, -2, 0, 0)$, and $(1, 0, -1, 2)$.

Let us denote these vectors by A, B, C. Let

$$B' = B - \frac{B \cdot A}{A \cdot A}\, A.$$

In other words, we subtract from B its projection along A. Then B' is perpendicular to A. We find

$$B' = \tfrac{1}{3}(4, -5, 0, 1).$$

Now we subtract from C its projection along A and B', and thus we let

$$C' = C - \frac{C \cdot A}{A \cdot A}\, A - \frac{C \cdot B'}{B' \cdot B'}\, B'.$$

Since A and B' are perpendicular, taking the scalar product of C' with A and B' shows that C' is perpendicular to both A and B'. We find

$$C' = \tfrac{1}{7}(-4, -2, -7, 6).$$

The vectors A, B', C' are non-zero and mutually perpendicular. They lie in the space generated by A, B, C. Hence they constitute an orthogonal basis for that space. If we wish an orthonormal basis, then we divide these vectors by their norm, and thus obtain

$$\frac{A}{\|A\|} = \frac{1}{\sqrt{3}}(1, 1, 0, 1),$$

$$\frac{B'}{\|B'\|} = \frac{1}{\sqrt{42}}(4, -5, 0, 1),$$

$$\frac{C'}{\|C'\|} = \frac{1}{\sqrt{105}}(-4, -2, -7, 6),$$

as an orthonormal basis.

Example 4. Find an orthogonal basis for the space of solutions of the linear equation

$$3x - 2y + z = 0.$$

First we find a basis, not necessarily orthogonal. For instance, we give z an arbitrary value, say $z = 1$. Thus we have to satisfy

$$3x - 2y = -1.$$

By inspection, we let $x = 1$, $y = 2$ or $x = 3$, $y = 5$, that is

$$A = (1, 2, 1) \qquad \text{and} \qquad B = (3, 5, 1).$$

Then it is easily verified that A, B are linearly independent. By Theorem 4.3 of Chapter 4, the space of solutions has dimension 2, so A, B form a basis of that space of solutions. To get an orthogonal basis, we start with A. Then we let

$$C = B - \text{projection of } B \text{ along } A$$

$$= B - \frac{B \cdot A}{A \cdot A}A$$

$$= \left(\frac{2}{3}, \frac{1}{3}, \frac{-4}{3}\right).$$

Then $\{A, C\}$ is an orthogonal basis of the space of solutions. It is sometimes convenient to get rid of the denominator. We may use

$$A = (1, 2, 1) \qquad \text{and} \qquad D = (2, 1, -4)$$

equally well for an orthogonal basis of that space. As a check, substitute back in the original equation to see that these vectors give solutions, and also verify that $A \cdot D = 0$, so that they are perpendicular to each other.

Example 5. Find an orthogonal basis for the space of solutions of the homogeneous equations

$$3x - 2y + z + \quad w = 0,$$
$$x + \quad y \quad + 2w = 0.$$

Let W be the space of solutions in $\mathbf{R}^4$. Then W is the space orthogonal to the two vectors

$$(3, -2, 1, 1) \quad \text{and} \quad (1, 1, 0, 2).$$

These are obviously linearly independent (by any number of arguments, you can prove at once that the matrix

$$\begin{pmatrix} -2 & 1 \\ 1 & 0 \end{pmatrix}$$

has rank 2, for instance). Hence

$$\dim W = 4 - 2 = 2.$$

Next we find a basis for the space of solutions. Let us put $w = 1$, and solve

$$3x - 2y + z = -1,$$
$$x + \quad y \quad = -2,$$

by ordinary elimination. If we put $y = 0$, then we get a solution with $x = -2$, and

$$z = -1 - 3x + 2y = 5.$$

If we put $y = 1$, then we get a solution with $x = -3$, and

$$z = -1 - 3x + 2y = 10.$$

Thus we get the two solutions

$$A = (-2, 0, 5, 1) \quad \text{and} \quad B = (-3, 1, 10, 1).$$

(As a check, substitute back in the original system of equations to see that no computational error has been made.) These two solutions are linearly independent, because for instance the matrix

$$\begin{pmatrix} -2 & 0 \\ -3 & 1 \end{pmatrix}$$

has rank 2. Hence $\{A, B\}$ is a basis for the space of solutions. To find an orthogonal basis, we orthogonalize B, to get

$$B' = B - \frac{B \cdot A}{A \cdot A} A = B - \frac{19}{10} A.$$

We can also clear denominators, and let $C = 10B'$, so

$$C = (-30, 10, 100, 10) - (-38, 0, 95, 19)$$
$$= (8, 10, 5, -9)$$

Then $\{A, C\}$ is an orthogonal basis for the space of solutions. (Again, check by substituting back in the system of equations, and also check perpendicularity by seeing directly that $A \cdot C = 0$.)

One can also find an orthogonal basis without guessing solutions by inspection or elimination at the start, as follows.

Example 6. Find a basis for the space of solutions of the equation

$$3x - 2y + z = 0.$$

The space of solutions is the space orthogonal to the vector $(3, -2, 1)$ and hence has dimension 2. There are of course many bases for this space. To find one, we first extend $(3, -2, 1) = A$ to a basis of $\mathbf{R}^3$. We do this by selecting vectors B, C such that A, B, C are linearly independent. For instance, take

$$B = (0, 1, 0)$$

and

$$C = (0, 0, 1).$$

Then A, B, C are linearly independent. To see this, we proceed as usual. If a, b, c are numbers such that

$$aA + bB + cC = 0,$$

then

$$3a \quad = 0,$$
$$-2a + b = 0,$$
$$a + c = 0.$$

This is easily solved to see that $a = b = c = 0$, so A, B, C are linearly independent. Now we must orthogonalize these vectors.
Let

$$B' = B - \frac{\langle B, A \rangle}{\langle A, A \rangle} A = \left(\frac{3}{7}, \frac{5}{7}, \frac{1}{7} \right),$$

$$C' = C - \frac{\langle C, A \rangle}{\langle A, A \rangle} A - \frac{\langle C, B' \rangle}{\langle B', B' \rangle} B'$$

$$= (0, 0, 1) - \tfrac{1}{14}(3, -2, 1) - \tfrac{1}{35}(3, 5, 1).$$

Then $\{B', C'\}$ is a basis for the space of solutions of the given equation. As you see, this procedure is slightly longer than the one used by guessing first, and involves two orthogonalizations rather than one as in Example 4.
In Theorem 2.1 we obtained an orthogonal basis for V by starting with an orthogonal basis for a subspace. Let us now look at the situation more symmetrically.

Theorem 2.3. *Let V be a vector space of dimension n, with a positive definite scalar product. Let $\{w_1, \ldots, w_r, u_1, \ldots, u_s\}$ be an orthogonal basis for V. Let W be the subspace generated by $w_1, \ldots, w_r$ and let U be the subspace generated by $u_1, \ldots, u_s$. Then $U = W^\perp$, or by symmetry, $W = U^\perp$. Hence for any subspace W of V we have the relation*

$$\dim W + \dim W^\perp = \dim V.$$

Proof. We shall prove that $W^\perp \subset U$ and $U \subset W^\perp$, so $W^\perp = U$.
First let $v \in W^\perp$. There exist numbers a_i $(i = 1, \ldots, r)$ and b_j $(j = 1, \ldots, s)$ such that

$$v = \sum_{i=1}^{r} a_i w_i + \sum_{j=1}^{s} b_j u_j.$$

Since v is perpendicular to all elements of W, we have for any $k = 1, \ldots, r$:

$$0 = v \cdot w_k = \sum a_i w_i \cdot w_k + \sum b_j u_j \cdot w_k$$
$$= a_k w_k \cdot w_k$$

because $w_i \cdot w_k = 0$ if $i \neq k$ and $u_j \cdot w_k = 0$ for all j. Since $w_k \cdot w_k \neq 0$ it follows that $a_k = 0$ for all $k = 1, \ldots, r$ so v is a linear combination of $u_1, \ldots, u_s$ and $v \in U$. Thus $W^\perp \subset U$.

Conversely, let $v \in U$, so v is a linear combination of $u_1, \ldots, u_s$. Since $\{w_1, \ldots, w_r, u_1, \ldots, u_s\}$ is an orthogonal basis of V it follows that each u_j is perpendicular to W so v itself is perpendicular to W, so $U \subset W^\perp$. Therefore we have proved that $U = W^\perp$.

Finally, Theorem 2.1 shows that the previous situation applies to any subspace W of V, and by the definition of dimension,

$$\dim V = r + s = \dim W + \dim W^\perp,$$

thus concluding the proof of the theorem.

Example 7. Consider $\mathbf{R}^3$. Let A, B be two linearly independent vectors in $\mathbf{R}^3$. Then the space of vectors which are perpendicular to both A and B is a 1-dimensional space. If $\{N\}$ is a basis for this space, any other basis for this space is of type $\{tN\}$, where t is a number $\neq 0$.

Again in $\mathbf{R}^3$, let N be a non-zero vector. The space of vectors perpendicular to N is a 2-dimensional space, i.e. a plane, passing through the origin O.

Remark. Theorem 2.3 gives a new proof of the fact that the row rank of a matrix is equal to its column rank. Indeed, let $A = (a_{ij})$ be an $m \times n$ matrix. Let S be the space of solutions of the equation $AX = O$, so $S = \operatorname{Ker} L_A$. By Theorem 3.2 of Chapter IV, we have

$$\dim S + \text{column rank} = n,$$

because the image of L_A is the space generated by the columns of A.

On the other hand, S is the space of vectors in $\mathbf{R}^n$ perpendicular to the rows of A, so if W is the row space then $S = W^\perp$. Therefore by Theorem 2.3 we get

$$\dim S + \text{row rank} = n.$$

This proves that row rank = column rank. In some ways, this is a more satisfying and conceptual proof of the relation than with the row and column operations that we used before.

We conclude this section by pointing out some useful notation. Let X, $Y \in \mathbf{R}^n$, and view X, Y as column vectors. Let $\langle \ , \ \rangle$ denote the standard scalar product on $\mathbf{R}^n$. Thus by definition

$$\langle X, Y \rangle = {}^t X Y.$$

Similarly, let A be an $n \times n$ matrix. Then

$$\langle X, AY \rangle = {}^tXAY = {}^t({}^tAX)Y = \langle {}^tAX, Y \rangle.$$

Thus we obtain the formula

$$\boxed{\langle X, AY \rangle = \langle {}^tAX, Y \rangle.}$$

The transpose of the matrix A corresponds to transposing A to ${}^tA$ from one side of the scalar product to the other. This notation is frequently used in applications, which is one of the reasons for mentioning it here.

Exercises VI, §2

1. Find orthonormal bases for the subspaces of $\mathbf{R}^3$ generated by the following vectors:
   (a) $(1, 1, -1)$ and $(1, 0, 1)$,
   (b) $(2, 1, 1)$ and $(1, 3, -1)$.

2. Find an orthonormal basis for the subspace of $\mathbf{R}^4$ generated by the vectors $(1, 2, 1, 0)$ and $(1, 2, 3, 1)$.

3. Find an orthonormal basis for the subspace of $\mathbf{R}^4$ generated by $(1, 1, 0, 0)$, $(1, -1, 1, 1)$, and $(-1, 0, 2, 1)$.

4. Find an orthogonal basis for the space of solutions of the following equations.
   (a) $2x + y - z = 0$ (b) $x - y + z = 0$
       $y + z = 0$
   (c) $4x + 7y - \pi z = 0$ (d) $x + y + z = 0$
       $2x - y + z = 0$ $x - y \quad\ = 0$
                                     $y + z = 0$

   In the next exercises, we consider the vector space of continuous functions on the interval $[0, 1]$. We define the scalar product of two such functions f, g by the rule

$$\langle f, g \rangle = \int_0^1 f(t)g(t)\ dt.$$

5. Let V be the subspace of functions generated by the two functions $f(t) = t$ and $g(t) = t^2$. Find an orthonormal basis for V.

6. Let V be the subspace generated by the three functions 1, t, t^2 (where 1 is the constant function). Find an orthonormal basis for V.

7. Let V be a finite dimensional vector space with a positive definite scalar product. Let W be a subspace. Show that

$$V = W + W^\perp \quad \text{and} \quad W \cap W^\perp = \{0\}.$$

In the terminology of the preceding chapter, this means that V is the direct sum of W and its orthogonal complement. [Use Theorem 2.3.]

8. In Exercise 7, show that $(W^{\perp})^{\perp} = W$. Why is this immediate from Theorem 2.3?

9. (a) Let V be the space of symmetric $n \times n$ matrices. For $A, B \in V$ define

$$\langle A, B \rangle = \text{tr}(AB),$$

where tr is the trace (sum of diagonal elements). Show that this satisfies all the properties of a positive definite scalar product. (You might already have done this as an exercise in a previous section.)

(b) Let W be the subspace of matrices A such that $\text{tr}(A) = 0$. What is the dimension of the orthogonal complement of W, relative to the scalar product in part (a)? Give an explicit basis for this orthogonal complement.

10. Let A be a symmetric $n \times n$ matrix. Let $X, Y \in \mathbf{R}^n$ be eigenvectors for A, that is suppose that there exist numbers a, b such that $AX = aX$ and $AY = bY$. Assume that $a \neq b$. Prove that X, Y are perpendicular.

VI, §3. Bilinear Maps and Matrices

Let U, V, W be vector spaces, and let

$$g: U \times V \to W$$

be a map. We say that g is **bilinear** if for each fixed $u \in U$ the map

$$v \mapsto g(u, v)$$

is linear, and for each fixed $v \in V$, the map

$$u \mapsto g(u, v)$$

is linear. The first condition written out reads

$$g(u, v_1 + v_2) = g(u, v_1) + g(u, v_2),$$
$$g(u, cv) = cg(u, v),$$

and similarly for the second condition on the other side.

Example. Let A be an $m \times n$ matrix, $A = (a_{ij})$. We can define a map

$$g_A: \mathbf{R}^m \times \mathbf{R}^n \to \mathbf{R}$$

by letting

$$g_A(X, Y) = {}^tXAY,$$

which written out looks like this:

$$(x_1, \ldots, x_m) \begin{pmatrix} a_{11} & \cdots & a_{1n} \\ \vdots & & \vdots \\ a_{m1} & \cdots & a_{mn} \end{pmatrix} \begin{pmatrix} y_1 \\ \vdots \\ y_n \end{pmatrix}.$$

Our vectors X and Y are supposed to be column vectors, so that ${}^tX$ is a row vector, as shown. Then ${}^tXA$ is a row vector, and ${}^tXAY$ is a 1×1 matrix, i.e. a number. Thus g_A maps pairs of vectors into the reals. Such a map g_A satisfies properties similar to those of a scalar product. If we fix X, then the map $Y \mapsto {}^tXAY$ is linear, and if we fix Y, then the map $X \mapsto {}^tXAY$ is also linear. In other words, say fixing X, we have

$$g_A(X, Y + Y') = g_A(X, Y) + g_A(X, Y'),$$

$$g_A(X, cY) = cg_A(X, Y),$$

and similarly on the other side. This is merely a reformulation of properties of multiplication of matrices, namely

$$^tXA(Y + Y') = {}^tXAY + {}^tXAY',$$

$$^tXA(cY) = c^tXAY.$$

It is convenient to write out the multiplication ${}^tXAY$ as a sum. Note that

$$j\text{-th component of } {}^tXA = \sum_{i=1}^{m} x_i a_{ij},$$

and thus

$$^tXAY = \sum_{j=1}^{n} \sum_{i=1}^{m} x_i a_{ij} y_j = \sum_{j=1}^{n} \sum_{i=1}^{m} a_{ij} x_i y_j.$$

Example. Let

$$A = \begin{pmatrix} 1 & 2 \\ 3 & -1 \end{pmatrix}$$

If $X = \begin{pmatrix} x_1 \\ x_2 \end{pmatrix}$ and $Y = \begin{pmatrix} y_1 \\ y_2 \end{pmatrix}$ then

$$^tXAY = x_1 y_1 + 2x_1 y_2 + 3x_2 y_1 - x_2 y_2.$$

Theorem 3.1. *Given a bilinear map $g: \mathbf{R}^m \times \mathbf{R}^n \to \mathbf{R}$, there exists a unique matrix A such that $g = g_A$, i.e. such that*

$$g(X, Y) = {}^t X A Y.$$

Proof. The statement of Theorem 3.1 is similar to the statement representing linear maps by matrices, and its proof is an extension of previous proofs. Remember that we used the standard bases for $\mathbf{R}^n$ to prove these previous results, and we used coordinates. We do the same here. Let $E^1, \dots, E^m$ be the standard unit vectors for $\mathbf{R}^m$, and let $U^1, \dots, U^n$ be the standard unit vectors for $\mathbf{R}^n$. We can then write any $X \in \mathbf{R}^m$ as

$$X = \sum_{i=1}^{m} x_i E^i$$

and any $Y \in \mathbf{R}^n$ as

$$Y = \sum_{j=1}^{n} y_j U^j.$$

Then

$$g(X, Y) = g(x_1 E^1 + \cdots + x_m E^m, y_1 U^1 + \cdots + y_n U^n).$$

Using the linearity on the left, we find

$$g(X, Y) = \sum_{i=1}^{m} x_i g(E^i, y_1 U^1 + \cdots + y_n U^n).$$

Using the linearity on the right, we find

$$g(X, Y) = \sum_{i=1}^{m} \sum_{j=1}^{n} x_i y_j g(E^i, U^j).$$

Let

$$a_{ij} = g(E^i, U^j).$$

Then we see that

$$g(X, Y) = \sum_{i=1}^{m} \sum_{j=1}^{n} a_{ij} x_i y_j,$$

which is precisely the expression we obtained for the product

$$^tXAY,$$

where A is the matrix (a_{ij}). This proves that $g = g_A$ for the choice of a_{ij} given above.

The uniqueness is also easy to see, and may be formulated as follows.

Uniqueness. *If A, B are $m \times n$ matrices such that for all vectors X, Y (of the appropriate dimension) we have*

$$^tXAY = {}^tXBY,$$

then $A = B$.

Proof. Since the above relation holds for *all* vectors X, Y, it holds in particular for the unit vectors. Thus we apply the relation when $X = E^i$ and $Y = U^j$. Then the rule for multiplication of matrices shows that

$$^tE^iAU^j = a_{ij} \qquad \text{and} \qquad {}^tE^iBU^j = b_{ij}.$$

Hence $a_{ij} = b_{ij}$ for all indices i, j. This shows that $A = B$.

Remark. Bilinear maps can be added and multiplied by scalars. The sum of two bilinear maps is again bilinear, and the product by a scalar is again bilinear. Hence bilinear maps form a vector space. Verify the rules

$$g_{A+B} = g_A + g_B \qquad \text{and} \qquad g_{cA} = cg_A.$$

Then Theorem 3.1 can be expressed by saying that the association

$$A \mapsto g_A$$

is an isomorphism between the space of $m \times n$ matrices, and the space of bilinear maps from $\mathbf{R}^m \times \mathbf{R}^n$ into $\mathbf{R}$.

Application to calculus. If you have had the calculus of several variables, you have associated with a function f of n variables the matrix of second partial derivatives

$$\left(\frac{\partial^2 f}{\partial x_i \, \partial x_j} \right).$$

This matrix may be viewed as the matrix associated with a bilinear map, which is called the **Hessian**. Note that this matrix is symmetric since it is proved that for sufficiently smooth functions, the partials commute, that is

$$\frac{\partial^2 f}{\partial x_i\, \partial x_j} = \frac{\partial^2 f}{\partial x_j\, \partial x_i}.$$

Exercises VI, §3

1. Let A be $n \times n$ matrix, and assume that A is symmetric, i.e. $A = {}^t A$. Let $\varphi_A \colon \mathbf{R}^n \times \mathbf{R}^n \to \mathbf{R}$ be its associated bilinear map. Show that

$$g_A(X, Y) = g_A(Y, X)$$

for all $X, Y \in \mathbf{R}^n$, and thus that g_A is a scalar product, i.e. satisfies conditions **SP 1**, **SP 2**, and **SP 3**.

2. Conversely, assume that A is an $n \times n$ matrix such that

$$g_A(X, Y) = g_A(Y, X)$$

for all X, Y. Show that A is symmetric.

3. Write out in full in terms of coordinates the expression for ${}^t X A Y$ when A is the following matrix, and X, Y are vectors of the corresponding dimension.

(a) $\begin{pmatrix} 2 & -3 \\ 4 & 1 \end{pmatrix}$　　　(b) $\begin{pmatrix} 4 & 1 \\ -2 & 5 \end{pmatrix}$　　　(c) $\begin{pmatrix} -5 & 2 \\ \pi & 7 \end{pmatrix}$

(d) $\begin{pmatrix} 1 & 2 & -1 \\ -3 & 1 & 4 \\ 2 & 5 & -1 \end{pmatrix}$　　(e) $\begin{pmatrix} -4 & 2 & 1 \\ 3 & 1 & 1 \\ 2 & 5 & 7 \end{pmatrix}$　　(f) $\begin{pmatrix} -\frac{1}{2} & 2 & -5 \\ 1 & \frac{2}{3} & 4 \\ -1 & 0 & 3 \end{pmatrix}$

CHAPTER VII

Determinants

We have worked with vectors for some time, and we have often felt the need of a method to determine when vectors are linearly independent. Up to now, the only method available to us was to solve a system of linear equations by the elimination method. In this chapter, we shall exhibit a very efficient computational method to solve linear equations, and determine when vectors are linearly independent.

The cases of 2×2 and 3×3 determinants will be carried out separately in full, because the general case of $n \times n$ determinants involves notation which adds to the difficulties of understanding determinants. Some proofs in the $n \times n$ case will be omitted.

VII, §1. Determinants of Order 2

Before stating the general properties of an arbitrary determinant, we shall consider a special case.

Let

$$A = \begin{pmatrix} a & b \\ c & d \end{pmatrix}$$

be a 2×2 matrix. We define its **determinant** to be $ad - bc$. Thus the determinant is a number. We denote it by

$$\begin{vmatrix} a & b \\ c & d \end{vmatrix} = ad - bc.$$

For example, the determinant of the matrix

$$\begin{pmatrix} 2 & 1 \\ 1 & 4 \end{pmatrix}$$

is equal to $2 \cdot 4 - 1 \cdot 1 = 7$. The determinant of

$$\begin{pmatrix} -2 & -3 \\ 4 & 5 \end{pmatrix}$$

is equal to $(-2) \cdot 5 - (-3) \cdot 4 = -10 + 12 = 2$.

The determinant can be viewed as a function of the matrix A. It can also be viewed as a function of its two columns. Let these be A^1 and A^2 as usual. Then we write the determinant as

$$D(A), \qquad \text{Det}(A), \qquad \text{or} \qquad D(A^1, A^2).$$

The following properties are easily verified by direct computation, which you should carry out completely.

Property 1. *As a function of the column vectors, the determinant is linear.*

This means: suppose for instance $A^1 = C + C'$ is a sum of two columns. Then

$$D(C + C', A^2) = D(C, A^2) + D(C', A^2).$$

If x is a number, then

$$D(xA^1, A^2) = xD(A^1, A^2).$$

A similar formula holds with respect to the second variable. The formula can be proved directly from the definition of the determinant. For instance, let b', d' be two numbers. Then

$$\text{Det}\begin{pmatrix} a & b + b' \\ c & d + d' \end{pmatrix} = a(d + d') - c(b + b')$$

$$= ad + ad' - cb - cb'$$

$$= ad - bc + ad' - b'c$$

$$= \text{Det}\begin{pmatrix} a & b \\ c & d \end{pmatrix} + \text{Det}\begin{pmatrix} a & b' \\ c & d' \end{pmatrix}.$$

Furthermore, if x is a number, then

$$\text{Det}\begin{pmatrix} xa & b \\ xc & d \end{pmatrix} = xad - xbc = x(ad - bc) = x \; \text{Det}\begin{pmatrix} a & b \\ c & d \end{pmatrix}.$$

In the terminology of Chapter VI, §4 we may say that *the determinant is bilinear.*

Property 2. *If the two columns are equal, then the determinant is equal to* 0.

This is immediate, since by hypothesis, the determinant is $ab - ab = 0$.

Property 3. *If I is the unit matrix, $I = (E^1, E^2)$, then*

$$D(I) = D(E^1, \; E^2) = 1.$$

Again this is immediate from the definition $ad - bc$.

Using only the above three properties we can prove others as follows.

If one adds a scalar multiple of one column to the other, then the value of the determinant does not change.

In other words, let x be a number. Then

$$D(A^1 + xA^2, \; A^2) = D(A^1, \; A^2).$$

The proof is immediate, namely:

$$D(A^1 + xA^2, A^2) = D(A^1, A^2) + xD(A^2, A^2) \quad \text{by linearity}$$
$$= D(A^1, A^2) \qquad\qquad\qquad \text{by Property 2.}$$

If the two columns are interchanged, then the determinant changes by a sign.

In other words, we have $D(A^2, A^1) = -D(A^1, A^2)$, or writing out the components,

$$\text{Det}\begin{pmatrix} a & b \\ c & d \end{pmatrix} = -\text{Det}\begin{pmatrix} b & a \\ d & c \end{pmatrix}.$$

One can of course prove this directly from the formula $ad - bc$. But let us also derive it from the property that if the two columns are equal, then the determinant is 0. We have:

$$0 = D(A^1 + A^2, A^1 + A^2) \quad \text{(because each variable is equal to } A^1 + A^2)$$
$$= D(A^1, A^1 + A^2) + D(A^2, A^1 + A^2) \quad \text{(by linearity in the first variable)}$$
$$= D(A^1, A^1) + D(A^1, A^2) + D(A^2, A^1) + D(A^2, A^2) \quad \text{(by linearity in the second variable)}$$
$$= D(A^1, A^2) + D(A^2, A^1).$$

Thus we see that $D(A^2, A^1) = -D(A^1, A^2)$. Observe that this proof used only the linearity in each variable, and the fact that

$$D(C, C) = 0$$

if C is a vector.

The determinant of A is equal to the determinant of its transpose, i.e. $D(A) = D(^tA)$.

Explicitly, we have

$$\text{Det}\begin{pmatrix} a & b \\ c & d \end{pmatrix} = \text{Det}\begin{pmatrix} a & c \\ b & d \end{pmatrix}.$$

This formula comes from the formula $ad - bc$ for the determinant.

The vectors A^1, A^2 are linearly dependent if and only if the determinant is 0.

We shall give a proof which follows the same pattern as in the generalization to higher dimensional spaces. First suppose that A^1, A^2 are linearly dependent, so there is a linear relation

$$xA^1 + yA^2 = 0$$

with not both x, y equal to 0. Say $x \neq 0$. Then we can solve

$$A^1 = zA^2 \qquad \text{where} \qquad z = -y/x.$$

Now we have

$$D(A^1, A^2) = D(zA^2, A^2) = zD(A^2, A^2) = 0$$

by using linearity and the property that if the two columns are equal, then the determinant is 0.

Conversely, suppose that A^1, A^2 are linearly independent. Then they must form a basis of $\mathbf{R}^2$, which has dimension 2. Hence we can express the unit vectors E^1, E^2 as linear combinations of A^1, A^2, say

$$E^1 = xA^1 + yA^2 \qquad \text{and} \qquad E^2 = zA^1 + wA^2,$$

where x, y, z, w are scalars. Now we have:

$$
\begin{aligned}
1 = D(E^1, E^2) &= D(xA^1 + yA^2, zA^1 + wA^2) \\
&= xzD(A^1, A^1) + xwD(A^1, A^2) + yzD(A^2, A^1) + ywD(A^2, A^2) \\
&= (xw - yz)D(A^1, A^2).
\end{aligned}
$$

Since this last product is 1, we must have $D(A^1, A^2) \neq 0$. This proves the desired assertion.

Finally we prove the uniqueness of the determinant, by a method which will work in general;

Theorem 1.1. *Let φ be a function of two vector variables A^1, $A^2 \in \mathbf{R}^2$ such that:*

φ is bilinear, that is φ is linear in each variable.

$\varphi(A^1, A^1) = 0$ for all $A^1 \in \mathbf{R}^2$.

$\varphi(E^1, E^2) = 1$ if E^1, E^2 are the standard unit vectors $\begin{pmatrix} 1 \\ 0 \end{pmatrix}, \begin{pmatrix} 0 \\ 1 \end{pmatrix}$.

Then $\varphi(A^1, A^2)$ is the determinant.

Proof. Write

$$A^1 = aE^1 + cE^2 \qquad \text{and} \qquad A^2 = bE^1 + dE^2.$$

Then

$$
\begin{aligned}
\varphi(A^1, A^2) &= \varphi(aE^1 + cE^2, bE^1 + dE^2) \\
&= ab\varphi(E^1, E^1) + ad\varphi(E^1, E^2) + cb\varphi(E^2, E^1) + cd\varphi(E^2, E^2) \\
&= ad\varphi(E^1, E^2) - bc\varphi(E^1, E^2) \\
&= (ad - bc)\varphi(E^1, E^2) \\
&= ad - bc.
\end{aligned}
$$

At each step we have used one of the properties proved previously. This proves the theorem.

VII, §2. 3 × 3 and $n \times n$ Determinants

We shall define determinants by induction, and give a formula for computing them at the same time. We deal with the 3×3 case.

We have already defined 2×2 determinants. Let

$$A = (a_{ij}) = \begin{pmatrix} a_{11} & a_{12} & a_{13} \\ a_{21} & a_{22} & a_{23} \\ a_{31} & a_{32} & a_{33} \end{pmatrix}$$

be a 3×3 matrix. We define its determinant according to the formula known as the **expansion by a row**, say the first row. That is, we define

$$(*) \qquad \mathrm{Det}(A) = a_{11} \begin{vmatrix} a_{22} & a_{23} \\ a_{32} & a_{33} \end{vmatrix} - a_{12} \begin{vmatrix} a_{21} & a_{23} \\ a_{31} & a_{33} \end{vmatrix} + a_{13} \begin{vmatrix} a_{21} & a_{22} \\ a_{31} & a_{32} \end{vmatrix}.$$

$$= \begin{vmatrix} a_{11} & a_{12} & a_{13} \\ a_{21} & a_{22} & a_{23} \\ a_{31} & a_{32} & a_{33} \end{vmatrix}.$$

We may describe this sum as follows. Let A_{ij} be the matrix obtained from A by deleting the i-th row and the j-th column. Then the sum expressing $\mathrm{Det}(A)$ can be written

$$a_{11}\,\mathrm{Det}(A_{11}) - a_{12}\,\mathrm{Det}(A_{12}) + a_{13}\,\mathrm{Det}(A_{13}).$$

In other words, each term consists of the product of an element of the first row and the determinant of the 2×2 matrix obtained by deleting the first row and the j-th column, and putting the appropriate sign to this term as shown.

Example 1. Let

$$A = \begin{pmatrix} 2 & 1 & 0 \\ 1 & 1 & 4 \\ -3 & 2 & 5 \end{pmatrix}.$$

Then

$$A_{11} = \begin{pmatrix} 1 & 4 \\ 2 & 5 \end{pmatrix}, \qquad A_{12} = \begin{pmatrix} 1 & 4 \\ -3 & 5 \end{pmatrix}, \qquad A_{13} = \begin{pmatrix} 1 & 1 \\ -3 & 2 \end{pmatrix}$$

and our formula for the determinant of A yields

$$\text{Det}(A) = 2\begin{vmatrix} 1 & 4 \\ 2 & 5 \end{vmatrix} - 1\begin{vmatrix} 1 & 4 \\ -3 & 5 \end{vmatrix} + 0\begin{vmatrix} 1 & 1 \\ -3 & 2 \end{vmatrix}$$

$$= 2(5 - 8) - 1(5 + 12) + 0$$

$$= -23.$$

The determinant of a 3×3 matrix can be written as

$$D(A) = \text{Det}(A) = D(A^1, A^2, A^3).$$

We use this last expression if we wish to consider the determinant as a function of the columns of A.

Furthermore, there is no particular reason why we selected the expansion according to the first row. We can also use the second row, and write a similar sum, namely:

$$-a_{21}\begin{vmatrix} a_{12} & a_{13} \\ a_{32} & a_{33} \end{vmatrix} + a_{22}\begin{vmatrix} a_{11} & a_{13} \\ a_{31} & a_{33} \end{vmatrix} - a_{23}\begin{vmatrix} a_{11} & a_{12} \\ a_{31} & a_{32} \end{vmatrix}$$

$$= -a_{21}\,\text{Det}(A_{21}) + a_{22}\,\text{Det}(A_{22}) - a_{23}\,\text{Det}(A_{23}).$$

Again, each term is the product of a_{2j}, the determinant of the 2×2 matrix obtained by deleting the second row and j-th column, and putting the appropriate sign in front of each term. This sign is determined according to the pattern:

$$\begin{pmatrix} + & - & + \\ - & + & - \\ + & - & + \end{pmatrix}.$$

One can see directly that the determinant can be expanded according to any row by multiplying out all the terms, and expanding the 2×2 determinants, thus obtaining the determinant as an alternating sum of six terms:

(**)

$$\text{Det}(A) = a_{11}a_{22}a_{33} - a_{11}a_{32}a_{23} - a_{12}a_{21}a_{33} + a_{12}a_{23}a_{31}$$
$$+ a_{13}a_{21}a_{32} - a_{13}a_{22}a_{31}.$$

We can also expand according to columns following the same principle. For instance, the expansion according to the first column:

$$a_{11}\begin{vmatrix} a_{22} & a_{23} \\ a_{32} & a_{33} \end{vmatrix} - a_{21}\begin{vmatrix} a_{12} & a_{13} \\ a_{32} & a_{33} \end{vmatrix} + a_{31}\begin{vmatrix} a_{12} & a_{13} \\ a_{22} & a_{23} \end{vmatrix}$$

yields precisely the same six terms as in (**).

In the case of 3×3 determinants, we therefore have the following result.

Theorem 2.1. *The determinant satisfies the rule for expansion according to rows and columns, and* $\text{Det}(A) = \text{Det}({}^t A)$. *In other words, the determinant of a matrix is equal to the determinant of its transpose.*

Example 2. Compute the determinant

$$\begin{vmatrix} 3 & 0 & 1 \\ 1 & 2 & 5 \\ -1 & 4 & 2 \end{vmatrix}$$

by expanding according to the second column.
   The determinant is equal to

$$2 \begin{vmatrix} 3 & 1 \\ -1 & 2 \end{vmatrix} - 4 \begin{vmatrix} 3 & 1 \\ 1 & 5 \end{vmatrix} = 2(6 - (-1)) - 4(15 - 1) = -42.$$

Note that the presence of a 0 in the second column eliminates one term in the expansion, since this term would be 0.
   We can also compute the above determinant by expanding according to the third column, namely the determinant is equal to

$$+ 1 \begin{vmatrix} 1 & 2 \\ -1 & 4 \end{vmatrix} - 5 \begin{vmatrix} 3 & 0 \\ -1 & 4 \end{vmatrix} + 2 \begin{vmatrix} 3 & 0 \\ 1 & 2 \end{vmatrix} = -42.$$

Next, let $A = (a_{ij})$ be an arbitrary $n \times n$ matrix. Let A_{ij} be the $(n - 1) \times (n - 1)$ matrix obtained by deleting the i-th row and j-th column from A.

$$A_{ij} = \begin{pmatrix} a_{11} & a_{12} & \cdots & a_{1j} & \cdots & a_{1n} \\ \vdots & \vdots & & | & & \vdots \\ \hline & & & -a_{ij}- & & \\ \vdots & \vdots & & | & & \vdots \\ a_{n1} & a_{n2} & \cdots & a_{nj} & \cdots & a_{nn} \end{pmatrix}$$

We give an expression for the determinant of an $n \times n$ matrix in terms of determinants of $(n - 1) \times (n - 1)$ matrices. Let i be an integer, $1 \leqq i \leqq n$. We define

$$D(A) = (-1)^{i+1} a_{i1} \, \text{Det}(A_{i1}) + \cdots + (-1)^{i+n} a_{in} \, \text{Det}(A_{in}).$$

This sum can be described in words. For each element of the i-th row, we have a contribution of one term in the sum. This term is equal to $+$ or $-$ the product of this element, times the determinant of the matrix obtained from A by deleting the i-th row and the corresponding column. The sign $+$ or $-$ is determined according to the chess-board pattern:

$$\begin{pmatrix} + & - & + & - & \cdots \\ - & + & - & + & \cdots \\ + & - & + & - & \cdots \end{pmatrix}.$$

This sum is called the **expansion of the determinant according to the i-th row**.

Using more complicated notation which we omit in this book, one can show that Theorem 2.1 is also valid in the $n \times n$ case. In particular, the determinant satisfies the rule of expansion according to the j-th column, for any j. Thus we have the expansion formula:

$$D(A) = (-1)^{1+j}a_{1j}D(A_{1j}) + \cdots + (-1)^{n+j}a_{nj}D(A_{nj}).$$

In practice, the computation of a determinant is always done by using an expansion according to some row or column.

We use the same notation for the determinant in the $n \times n$ case that we used in the 2×2 or 3×3 cases, namely

$$|A| = D(A) = \text{Det}(A) = D(A^1, \ldots, A^n).$$

The notation $D(A^1, \ldots, A^n)$ is especially suited to denote the determinant as a function of the columns, for instance to state the next theorem.

Theorem 2.2. *The determinant satisfies the following properties:*

1. *As a function of each column vector, the determinant is linear, i.e. if the j-th column A^j is equal to a sum of two column vectors, say $A^j = C + C'$, then*

$$D(A^1, \ldots, C + C', \ldots, A^n)$$
$$= D(A^1, \ldots, C, \ldots, A^n) + D(A^1, \ldots, C', \ldots, A^n).$$

Furthermore, if x is a number, then

$$D(A^1, \ldots, xA^j, \ldots, A^n) = xD(A^1, \ldots, A^j, \ldots, A^n).$$

2. *If two columns are equal, i.e. if $A^j = A^k$, with $j \neq k$, then the determinant $D(A)$ is equal to 0.*

3. *If I is the unit matrix, then $D(I) = 1$.*

The proof requires more complicated notation and will be omitted. It can be carried out by induction, and from the explicit formula giving the expansion of the determinant.

As an example we give the proof in the case of 3×3 determinants. The proof is by direct computation. Suppose say that the first column is a sum of two columns:

$$A^1 = B + C, \qquad \text{that is,} \qquad \begin{pmatrix} a_{11} \\ a_{21} \\ a_{31} \end{pmatrix} = \begin{pmatrix} b_1 \\ b_2 \\ b_3 \end{pmatrix} + \begin{pmatrix} c_1 \\ c_2 \\ c_3 \end{pmatrix}.$$

Substituting in each term of $(*)$, we see that each term splits into a sum of two terms corresponding to B and C. For instance,

$$a_{11} \begin{vmatrix} a_{22} & a_{23} \\ a_{32} & a_{33} \end{vmatrix} = b_1 \begin{vmatrix} a_{22} & a_{23} \\ a_{32} & a_{33} \end{vmatrix} + c_1 \begin{vmatrix} a_{22} & a_{23} \\ a_{32} & a_{33} \end{vmatrix},$$

$$a_{12} \begin{vmatrix} b_2 + c_2 & a_{23} \\ b_3 + c_3 & a_{33} \end{vmatrix} = a_{12} \begin{vmatrix} b_2 & a_{23} \\ b_3 & a_{33} \end{vmatrix} + a_{12} \begin{vmatrix} c_2 & a_{23} \\ c_3 & a_{33} \end{vmatrix},$$

and similarly for the third term. The proof with respect to the other column is analogous. Furthermore, if x is a number, then

$$\text{Det}(xA^1, A^2, A^3) = xa_{11} \begin{vmatrix} a_{22} & a_{23} \\ a_{32} & a_{33} \end{vmatrix} - a_{12} \begin{vmatrix} xa_{21} & a_{23} \\ xa_{31} & a_{33} \end{vmatrix} + a_{13} \begin{vmatrix} xa_{21} & a_{23} \\ xa_{31} & a_{32} \end{vmatrix}$$

$$= x \, \text{Det}(A^1, A^2, A^3)$$

Next, suppose that two columns are equal, for instance the first and second, so $A^1 = A^2$. Thus

$$a_{11} = a_{12}, \qquad a_{21} = a_{22}, \qquad a_{31} = a_{32}.$$

Then again you can see directly that terms will cancel to make the determinant equal to 0.

Finally, if I is the unit 3×3 matrix, then $\text{Det}(I) = 1$ by using the expansion according to either rows or columns, because in such an expansion all but one term are equal to 0, and this single term is equal to 1 times the determinant of the unit 2×2 matrix, which is also equal to 1.

A function of several variables which is linear in each variable, i.e. which satisfies the first property of determinants, is called **multilinear**. A function which satisfies the second property is called **alternating**.

To compute determinants efficiently, we need additional properties which will be deduced simply from properties **1, 2, 3** of Theorem 2.2.

4. *Let j, k be integers with $1 \leq j \leq n$ and $1 \leq k \leq n$, and $j \neq k$. If the j-th column and k-th column are interchanged, then the determinant changes by a sign.*

Proof. In the matrix A be replace the j-th column and the k-th column by $A^j + A^k$. We obtain a matrix with two equal columns, so by property **2**, the determinant is 0. We expand by property **1** to get:

$$0 = D(\ldots,A^j + A^k,\ldots,A^j + A^k,\ldots)$$
$$= D(\ldots,A^j,\ldots,A^j,\ldots) + D(\ldots,A^j,\ldots,A^k,\ldots)$$
$$+ D(\ldots,A^k,\ldots,A^j,\ldots) + D(\ldots,A^k,\ldots,A^k,\ldots).$$

Using property **2** again, we see that two of these four terms are equal to 0, and hence that

$$0 = D(\ldots,A^j,\ldots,A^k,\ldots) + D(\ldots,A^k,\ldots,A^j,\ldots).$$

In this last sum, one term must be equal to minus the other, thus proving property **4**.

5. *If one adds a scalar multiple of one column to another then the value of the determinant does not change.*

Proof. Consider two different columns, say the k-th and j-th columns A^k and A^j with $k \neq j$. Let x be a scalar. We add xA^j to A^k. By property **1**, the determinant becomes

$$D(\ldots,A^k + xA^j,\ldots) = D(\ldots,A^k,\ldots) + D(\ldots,xA^j,\ldots)$$
$$\qquad\quad \uparrow \qquad\qquad\qquad \uparrow \qquad\qquad\quad \uparrow$$
$$\qquad\quad k \qquad\qquad\qquad k \qquad\qquad\quad k$$

(the k points to the k-th column). In both terms on the right, the indicated column occurs in the k-th place. But $D(\ldots,A^k,\ldots)$ is simply $D(A)$. Furthermore,

$$D(\ldots,xA^j,\ldots) = xD(\ldots,A^j,\ldots).$$
$$\quad \uparrow \qquad\qquad\qquad\quad \uparrow$$
$$\quad k \qquad\qquad\qquad\quad k$$

Since $k \neq j$, the determinant on the right has two equal columns, because A^j occurs in the k-th place and also in the j-th place. Hence it is equal to 0. Hence

$$D(\ldots,A^k + xA^j,\ldots) = D(\ldots,A^k,\ldots),$$

thereby proving our property **5**.

Since the determinant of a matrix is equal to the determinant of its transpose, that is $\text{Det}(A) = \text{Det}({}^t A)$, we obtain the following general fact:

*All the properties stated above for rows or column operations are valid for **both** row and column operations.*

For instance, if a scalar multiple of one row is added to another row, then the value of the determinant does not change.

With the above means at our disposal, we can now compute 3×3 determinants very efficiently. In doing so, we apply the operations described in property **5**. We try to make as many entries in the matrix A equal to 0. We try especially to make all but one element of a column (or row) equal to 0, and then expand according to that column (or row). The expansion will contain only one term, and reduces our computation to a 2×2 determinant.

Example 3. Compute the determinant

$$\begin{vmatrix} 3 & 0 & 1 \\ 1 & 2 & 5 \\ -1 & 4 & 2 \end{vmatrix}.$$

We already have 0 in the first row. We subtract twice the second row from the third row. Our determinant is then equal to

$$\begin{vmatrix} 3 & 0 & 1 \\ 1 & 2 & 5 \\ -3 & 0 & -8 \end{vmatrix}.$$

We expand according to the second column. The expansion has only one term $\neq 0$, with a $+$ sign, and that is:

$$2 \begin{vmatrix} 3 & 1 \\ -3 & -8 \end{vmatrix}.$$

The 2×2 determinant can be evaluated by our definition $ad - bc$, and we find $2(-24 - (-3)) = -42$.

Similarly, we reduce the computation of a 4×4 determinant to that of 3×3 determinants, and then 2×2 determinants.

Example 4. We wish to compute the determinant

$$\begin{vmatrix} 1 & 3 & 1 & 1 \\ 2 & 1 & 5 & 2 \\ 1 & -1 & 2 & 3 \\ 4 & 1 & -3 & 7 \end{vmatrix}.$$

We add the third row to the second row, and then add the third row to the fourth row. This yields

$$
\begin{vmatrix}
1 & 3 & 1 & 1 \\
3 & 0 & 7 & 5 \\
1 & -1 & 2 & 3 \\
4 & 1 & -3 & 7
\end{vmatrix}
=
\begin{vmatrix}
1 & 3 & 1 & 1 \\
3 & 0 & 7 & 5 \\
1 & -1 & 2 & 3 \\
5 & 0 & -1 & 10
\end{vmatrix}.
$$

We then add three times the third row to the first row and get

$$
\begin{vmatrix}
4 & 0 & 7 & 10 \\
3 & 0 & 7 & 5 \\
1 & -1 & 2 & 3 \\
5 & 0 & -1 & 10
\end{vmatrix}
$$

which we expand according to the second column. There is only one term, namely

$$
\begin{vmatrix}
4 & 7 & 10 \\
3 & 7 & 5 \\
5 & -1 & 10
\end{vmatrix}.
$$

We subtract twice the second row from the first row, and then from the third row, yielding

$$
\begin{vmatrix}
-2 & -7 & 0 \\
3 & 7 & 5 \\
-1 & -15 & 0
\end{vmatrix}.
$$

which we expand according to the third column, and get

$$
-5(30 - 7) = -5(23) = -115.
$$

Exercises VII, §2

1. Compute the following determinants.

(a) $\begin{vmatrix} 2 & 1 & 2 \\ 0 & 3 & -1 \\ 4 & 1 & 1 \end{vmatrix}$ (b) $\begin{vmatrix} 3 & -1 & 5 \\ -1 & 2 & 1 \\ -2 & 4 & 3 \end{vmatrix}$ (c) $\begin{vmatrix} 2 & 4 & 3 \\ -1 & 3 & 0 \\ 0 & 2 & 1 \end{vmatrix}$

(d) $\begin{vmatrix} 1 & 2 & -1 \\ 0 & 2 & -1 \\ 0 & 2 & 7 \end{vmatrix}$ (e) $\begin{vmatrix} -1 & 5 & 3 \\ 4 & 0 & 0 \\ 2 & 7 & 8 \end{vmatrix}$ (f) $\begin{vmatrix} 3 & 1 & 2 \\ 4 & 5 & 1 \\ -1 & 2 & -3 \end{vmatrix}$

2. Compute the following determinants.

(a) $\begin{vmatrix} 1 & 1 & -2 & 4 \\ 0 & 1 & 1 & 3 \\ 2 & -1 & 1 & 0 \\ 3 & 1 & 2 & 5 \end{vmatrix}$
(b) $\begin{vmatrix} -1 & 1 & 2 & 0 \\ 0 & 3 & 2 & 1 \\ 0 & 4 & 1 & 2 \\ 3 & 1 & 5 & 7 \end{vmatrix}$
(c) $\begin{vmatrix} 3 & 1 & 1 \\ 2 & 5 & 5 \\ 8 & 7 & 7 \end{vmatrix}$

(d) $\begin{vmatrix} 4 & -9 & 2 \\ 4 & -9 & 2 \\ 3 & 1 & 0 \end{vmatrix}$
(e) $\begin{vmatrix} 4 & -1 & 1 \\ 2 & 0 & 0 \\ 1 & 5 & 7 \end{vmatrix}$
(f) $\begin{vmatrix} 2 & 0 & 0 \\ 1 & 1 & 0 \\ 8 & 5 & 7 \end{vmatrix}$

(g) $\begin{vmatrix} 4 & 0 & 0 \\ 0 & 1 & 0 \\ 0 & 0 & 27 \end{vmatrix}$
(h) $\begin{vmatrix} 5 & 0 & 0 \\ 0 & 3 & 0 \\ 0 & 0 & 9 \end{vmatrix}$
(i) $\begin{vmatrix} 2 & -1 & 4 \\ 3 & 1 & 5 \\ 1 & 2 & 3 \end{vmatrix}$

3. In general, what is the determinant of a diagonal matrix

$$\begin{vmatrix} a_{11} & 0 & 0 & \cdots & 0 \\ 0 & a_{22} & 0 & \cdots & 0 \\ \vdots & \vdots & & & \vdots \\ 0 & 0 & & \ddots & 0 \\ 0 & 0 & 0 & \cdots & a_{nn} \end{vmatrix}?$$

4. Compute the determinant $\begin{vmatrix} \cos\theta & -\sin\theta \\ \sin\theta & \cos\theta \end{vmatrix}$.

5. (a) Let x_1, x_2, x_3 be numbers. Show that

$$\begin{vmatrix} 1 & x_1 & x_1^2 \\ 1 & x_2 & x_2^2 \\ 1 & x_3 & x_3^2 \end{vmatrix} = (x_2 - x_1)(x_3 - x_1)(x_3 - x_2).$$

*(b) If $x_1,\ldots,x_n$ are numbers, then show by induction that

$$\begin{vmatrix} 1 & x_1 & \cdots & x_1^{n-1} \\ 1 & x_2 & \cdots & x_2^{n-1} \\ 1 & x_n & \cdots & x_n^{n-1} \end{vmatrix} = \prod_{i<j}(x_j - x_i),$$

the symbol on the right meaning that it is the product of all terms $x_j - x_i$ with $i < j$ and i, j integers from 1 to n. This determinant is called the **Vandermonde** determinant V_n. To do the induction easily, multiply each column by x_1 and subtract it from the next column on the right, starting from the right-hand side. You will find that

$$V_n = (x_n - x_1)\cdots(x_2 - x_1)V_{n-1}.$$

6. Find the determinants of the following matrices.

(a) $\begin{pmatrix} 1 & 2 & 5 \\ 0 & 1 & 7 \\ 0 & 0 & 3 \end{pmatrix}$ (b) $\begin{pmatrix} -1 & 5 & 20 \\ 0 & 4 & 8 \\ 0 & 0 & 6 \end{pmatrix}$

(c) $\begin{pmatrix} 2 & -6 & 9 \\ 0 & 1 & 4 \\ 0 & 0 & 8 \end{pmatrix}$ (d) $\begin{pmatrix} -7 & 98 & 54 \\ 0 & 2 & 46 \\ 0 & 0 & -1 \end{pmatrix}$

(e) $\begin{pmatrix} 1 & 4 & 6 \\ 0 & 0 & 1 \\ 0 & 0 & 8 \end{pmatrix}$ (f) $\begin{pmatrix} 4 & 0 & 0 \\ -5 & 2 & 0 \\ 79 & 54 & 1 \end{pmatrix}$

(g) $\begin{pmatrix} 1 & 5 & 2 & 3 \\ 0 & 2 & 7 & 6 \\ 0 & 0 & 4 & 1 \\ 0 & 0 & 0 & 5 \end{pmatrix}$ (h) $\begin{pmatrix} -5 & 0 & 0 & 0 \\ 7 & 2 & 0 & 0 \\ -9 & 4 & 1 & 0 \\ 96 & 2 & 3 & 1 \end{pmatrix}$

(i) Let A be a triangular $n \times n$ matrix, say a matrix such that all components below the diagonal are equal to 0.

$$A = \begin{pmatrix} a_{11} & & & & \\ 0 & a_{22} & & * & \\ 0 & 0 & & & \\ \vdots & \vdots & & \ddots & \vdots \\ 0 & 0 & & \cdots & a_{nn} \end{pmatrix}.$$

What is $D(A)$?

7. If $a(t)$, $b(t)$, $c(t)$, $d(t)$ are functions of t, one can form the determinant

$$\begin{vmatrix} a(t) & b(t) \\ c(t) & d(t) \end{vmatrix},$$

just as with numbers. Write out in full the determinant

$$\begin{vmatrix} \sin t & \cos t \\ -\cos t & \sin t \end{vmatrix}.$$

8. Write out in full the determinant

$$\begin{vmatrix} t+1 & t-1 \\ t & 2t+5 \end{vmatrix}.$$

9. (*With calculus*) Let $f(t)$, $g(t)$ be two functions having derivatives of all orders. Let $\varphi(t)$ be the function obtained by taking the determinant

$$\varphi(t) = \begin{vmatrix} f(t) & g(t) \\ f'(t) & g'(t) \end{vmatrix}.$$

Show that

$$\varphi'(t) = \begin{vmatrix} f(t) & g(t) \\ f''(t) & g''(t) \end{vmatrix},$$

i.e. the derivative is obtained by taking the derivative of the bottom row.

10. (*With calculus*). Let

$$A(t) = \begin{pmatrix} b_1(t) & c_1(t) \\ b_2(t) & c_2(t) \end{pmatrix}$$

be a 2×2 matrix of differentiable functions. Let $B(t)$ and $C(t)$ be its column vectors. Let

$$\varphi(t) = \text{Det}(A(t)).$$

Show that

$$\varphi'(t) = D(B'(t), C(t)) + D(B(t), C'(t)).$$

11. Let c be a number and let A be a 3×3 matrix. Show that

$$D(cA) = c^3 D(A).$$

12. Let c be a number and let A be an $n \times n$ matrix. Show that

$$D(cA) = c^n D(A).$$

13. Let $c_1,\dots,c_n$ be numbers. How do the determinants differ:

$$D(c_1 A^1,\dots,c_n A^n) \quad \text{and} \quad D(A^1,\dots,A^n)?$$

14. Write down explicitly the expansion of a 4×4 determinant according to the first row and according to the first column.

VII, §3. The Rank of a Matrix and Subdeterminants

In this section we give a test for linear independence by using determinants.

Theorem 3.1. *Let $A^1,\dots,A^n$ be column vectors of dimension n. They are linearly dependent if and only if*

$$D(A^1,\dots,A^n) = 0.$$

Proof. Suppose $A^1,\ldots,A^n$ are linearly dependent, so there exists a relation

$$x_1 A^1 + \cdots + x_n A^n = O$$

with numbers $x_1,\ldots,x_n$ not all 0. Say $x_j \neq 0$. Subtracting and dividing by x_j, we can find numbers c_k with $k \neq j$ such that

$$A^j = \sum_{k \neq j} c_k A^k.$$

Thus

$$D(A) = D\left(A^1, \ldots, \sum_{k \neq j} c_k A^k, \ldots, A^n\right)$$

$$= \sum_{k \neq j} D(A^1, \ldots, A^k, \ldots, A^n)$$

where A^k occurs in the j-th place. But A^k also occurs in the k-th place, and $k \neq j$. Hence the determinant is equal to 0 by property **2**. This concludes the proof of the first part.

As to the converse, we recall that a matrix is row equivalent to an echelon matrix. Suppose that $A^1,\ldots,A^n$ are linearly independent. Then the matrix

$$A = (A^1, \ldots, A^n)$$

is row equivalent to a triangular matrix. Indeed, it is row equivalent to a matrix B in echelon form

$$\begin{pmatrix} b_{11} & b_{12} & \cdots & b_{1n} \\ 0 & b_{22} & \cdots & b_{2n} \\ \vdots & \vdots & & \vdots \\ 0 & 0 & \cdots & b_{nn} \end{pmatrix}$$

and the operations of row equivalence do not change the property of rows or columns being linearly independent. Hence all the diagonal elements $b_{11},\ldots,b_{nn}$ are $\neq 0$. The determinant of this matrix is the product

$$b_{11} \cdots b_{nn} \neq 0$$

by the rule of expansion according to columns. Under the operations of row equivalence, the property of the determinant being $\neq 0$ does not change, because row equivalences involve multiplying a row by a non-zero scalar which multiplies the determinant by this scalar; or interchanging rows, which multiplies the determinant by -1; or adding a multiple of one row to another, which does not change the value of the determinant. Since $\text{Det}(B) \neq 0$ it follows that $\text{Det}(A) \neq 0$. This concludes the proof.

Corollary 3.2. *If $A^1, \ldots, A^n$ are column vectors of $\mathbf{R}^n$ such that $D(A^1, \ldots, A^n) \neq 0$, and if B is a column vector, then there exist numbers $x_1, \ldots, x_n$ such that*

$$x_1 A^1 + \cdots + x_n A^n = B.$$

These numbers are uniquely determined by B.

Proof. According to the theorem, $A^1, \ldots, A^n$ are linearly independent, and hence form a basis of $\mathbf{R}^n$. Hence any vector of $\mathbf{R}^n$ can be written as a linear combination of $A^1, \ldots, A^n$. Since $A^1, \ldots, A^n$ are linearly independent, the numbers $x_1, \ldots, x_n$ are unique.

In terms of linear equations, this corollary shows:

If a system of n linear equations in n unknowns has a matrix of coefficients whose determinant is not 0, then this system has a unique solution.

Since determinants can be used to test linear independence, they can be used to determine the rank of a matrix.

Example 1. Let

$$A = \begin{pmatrix} 3 & 1 & 2 & 5 \\ 1 & 2 & -1 & 2 \\ 1 & 1 & 0 & 1 \end{pmatrix}.$$

This is a 3×4 matrix. Its rank is at most 3. If we can find three linearly independent columns, then we know that its rank is exactly 3. But the determinant

$$\begin{vmatrix} 1 & 2 & 5 \\ 2 & -1 & 2 \\ 1 & 0 & 1 \end{vmatrix} = 4$$

is not equal to 0. Hence rank $A = 3$.

It may be that in a 3×4 matrix, some determinant of a 3×3 submatrix is 0, but the 3×4 matrix has rank 3. For instance, let

$$B = \begin{pmatrix} 3 & 1 & 2 & 5 \\ 1 & 2 & -1 & 2 \\ 4 & 3 & 1 & 1 \end{pmatrix}.$$

The determinant of the first three columns

$$\begin{vmatrix} 3 & 1 & 2 \\ 1 & 2 & -1 \\ 4 & 3 & 1 \end{vmatrix}$$

is equal to 0 (in fact, the last row is the sum of the first two rows). But the determinant

$$\begin{vmatrix} 1 & 2 & 5 \\ 2 & -1 & 2 \\ 3 & 1 & 1 \end{vmatrix}$$

is not zero (what is it?) so that again the rank of B is equal to 3.

If the rank of a 3×4 matrix

$$C = \begin{pmatrix} c_{11} & c_{12} & c_{13} & c_{14} \\ c_{21} & c_{22} & c_{23} & c_{24} \\ c_{31} & c_{32} & c_{33} & c_{34} \end{pmatrix}$$

is 2 or less, then the determinant of *every* 3×3 submatrix must be 0, otherwise we could argue as above to get three linearly independent columns. We note that there are four such subdeterminants, obtained by eliminating successively any one of the four columns. Conversely, if every such subdeterminant of every 3×3 submatrix is equal to 0, then it is easy to see that the rank is at most 2. Because if the rank were equal to 3, then there would be three linearly independent columns, and their determinant would not be 0. Thus we can compute such subdeterminants to get an estimate on the rank, and then use trial and error, and some judgment, to get the exact rank.

Example 2. Let

$$C = \begin{pmatrix} 3 & 1 & 2 & 5 \\ 1 & 2 & -1 & 2 \\ 4 & 3 & 1 & 7 \end{pmatrix}.$$

If we compute every 3×3 subdeterminant, we shall find 0. Hence the rank of C is at most equal to 2. However, the first two rows are linearly independent, for instance because the determinant

$$\begin{vmatrix} 3 & 1 \\ 1 & 2 \end{vmatrix}$$

is not equal to 0. It is the determinant of the first two columns of the
2×4 matrix

$$\begin{pmatrix} 3 & 1 & 2 & 5 \\ 1 & 2 & -1 & 2 \end{pmatrix}.$$

Hence the rank is equal to 2.

Of course, if we notice that the last row of C is equal to the sum of
the first two, then we see at once that the rank is ≤ 2.

Exercises VII, §3

Compute the ranks of the following matrices.

1. $\begin{pmatrix} 2 & 3 & 5 & 1 \\ 1 & -1 & 2 & 1 \end{pmatrix}$

2. $\begin{pmatrix} 3 & 5 & 1 & 4 \\ 2 & -1 & 1 & 1 \\ 5 & 4 & 2 & 5 \end{pmatrix}$

3. $\begin{pmatrix} 3 & 5 & 1 & 4 \\ 2 & -1 & 1 & 1 \\ 8 & 9 & 3 & 9 \end{pmatrix}$

4. $\begin{pmatrix} 3 & 5 & 1 & 4 \\ 2 & -1 & 1 & 1 \\ 7 & 1 & 2 & 5 \end{pmatrix}$

5. $\begin{pmatrix} -1 & 1 & 6 & 5 \\ 1 & 1 & 2 & 3 \\ -1 & 2 & 5 & 4 \\ 2 & 1 & 0 & 1 \end{pmatrix}$

6. $\begin{pmatrix} 2 & 1 & 6 & 6 \\ 3 & 1 & 1 & -1 \\ 5 & 2 & 7 & 5 \\ -2 & 4 & 3 & 2 \end{pmatrix}$

7. $\begin{pmatrix} 2 & 1 & 6 & 6 \\ 3 & 1 & 1 & -1 \\ 5 & 2 & 7 & 5 \\ 8 & 3 & 8 & 4 \end{pmatrix}$

8. $\begin{pmatrix} 3 & 1 & 1 & -1 \\ -2 & 4 & 3 & 2 \\ -1 & 9 & 7 & 3 \\ 7 & 4 & 2 & 1 \end{pmatrix}$

9. *(With calculus).* Let $\alpha_1, \ldots, \alpha_n$ be distinct numbers, $\neq 0$. Show that the functions

$$e^{\alpha_1 t}, \ldots, e^{\alpha_n t}$$

are linearly independent over the numbers. [*Hint:* Suppose we have a linear
relation

$$c_1 e^{\alpha_1 t} + \cdots + c_n e^{\alpha_n t} = 0$$

with constants c_i, valid for all t. If not all c_i are 0, without loss of generality,
we may assume that none of them is 0. Differentiate the above relation $n - 1$
times. You get a system of linear equations. The determinant of its
coefficients must be zero. (Why?) Get a contradiction from this.]

VII, §4. Cramer's Rule

The properties of determinants can be used to prove a well-known rule
used in solving linear equations.

Theorem 4.1. *Let* $A^1,\ldots,A^n$ *be column vectors such that*

$$D(A^1,\ldots,A^n) \neq 0.$$

Let B be a column vector. If $x_1,\ldots,x_n$ *are numbers such that*

$$x_1 A^1 + \cdots + x_n A^n = B,$$

then for each $j = 1,\ldots,n$ *we have*

$$x_j = \frac{D(A^1,\ldots,B,\ldots,A^n)}{D(A^1,\ldots,A_n)}$$

where B occurs in the j-th column instead of A^j. *In other words,*

$$x_j = \frac{\begin{vmatrix} a_{11} & \cdots & b_1 & \cdots & a_{1n} \\ a_{21} & \cdots & b_2 & \cdots & a_{2b} \\ \vdots & & \vdots & & \vdots \\ a_{n1} & \cdots & b_n & \cdots & a_{nn} \end{vmatrix}}{\begin{vmatrix} a_{11} & \cdots & a_{1j} & \cdots & a_{1n} \\ a_{21} & \cdots & a_{2j} & \cdots & a_{2n} \\ \vdots & & \vdots & & \vdots \\ a_{n1} & \cdots & a_{nj} & \cdots & a_{nn} \end{vmatrix}}$$

(The numerator is obtained from A by replacing the j-th column A^j *by B. The denominator is the determinant of the matrix A.)*

Theorem 4.1 gives us an explicit way of finding the coordinates of B with respect to $A^1,\ldots,A^n$. In the language of linear equations, Theorem 4.1 allows us to solve explicitly in terms of determinants the system of n linear equations in n unknowns:

$$x_1 a_{11} + \cdots + x_n a_{1n} = b_1$$
$$\vdots \qquad\qquad \vdots \quad \vdots$$
$$x_1 a_{n1} + \cdots + x_n a_{nn} = b_n.$$

We prove Theorem 4.1.

Let B be written as in the statement of the theorem, and consider the determinant of the matrix obtained by replacing the j-th column A by B. Then

$$D(A^1,\ldots,B,\ldots,A^n) = D(A^1,\ldots,x_1 A_1^1 + \cdots + x_n A_n^1,\ldots,A^n).$$

We use property **1** and obtain a sum:

$$D(A^1,\ldots,x_1A^1,\ldots,A^n) + \cdots + D(A^1,\ldots,x_jA^j,\ldots,A^n)$$
$$+ \cdots + D(A^1,\ldots,x_nA^n,\ldots,A^n),$$

which by property **1** again, is equal to

$$x_1D(A^1,\ldots,A^1,\ldots,A^n) + \cdots + x_jD(A^1,\ldots,A^n)$$
$$+ \cdots + x_nD(A^1,\ldots,A^n,\ldots,A^n).$$

In every term of this sum except the j-th term, two column vectors are equal. Hence every term except the j-th term is equal to 0, by property **2**. The j-th term is equal to

$$x_jD(A^1,\ldots,A^n),$$

and is therefore equal to the determinant we started with, namely $D(A^1,\ldots,B,\ldots,A^n)$. We can solve for x_j, and obtain precisely the expression given in the statement of the theorem.

The rule of Theorem 4.1 giving us the solution to the system of linear equations by means of determinants, is known as **Cramer's rule**.

Example. Solve the system of linear equations:

$$3x + 2y + 4z = 1,$$
$$2x - y + z = 0,$$
$$x + 2y + 3z = 1.$$

We have:

$$x = \frac{\begin{vmatrix} 1 & 2 & 4 \\ 0 & -1 & 1 \\ 1 & 2 & 3 \end{vmatrix}}{\begin{vmatrix} 3 & 2 & 4 \\ 2 & -1 & 1 \\ 1 & 2 & 3 \end{vmatrix}}, \quad y = \frac{\begin{vmatrix} 3 & 1 & 4 \\ 2 & 0 & 1 \\ 1 & 1 & 3 \end{vmatrix}}{\begin{vmatrix} 3 & 2 & 4 \\ 2 & -1 & 1 \\ 1 & 2 & 3 \end{vmatrix}}, \quad z = \frac{\begin{vmatrix} 3 & 2 & 1 \\ 2 & -1 & 0 \\ 1 & 2 & 1 \end{vmatrix}}{\begin{vmatrix} 3 & 2 & 4 \\ 2 & -1 & 1 \\ 1 & 2 & 3 \end{vmatrix}}.$$

Observe how the column

$$B = \begin{pmatrix} 1 \\ 0 \\ 1 \end{pmatrix}$$

shifts from the first column when solving for x, to the second column when solving for y, to the third column when solving for z. The denominator in all three expressions is the same, namely it is the determinant of the matrix of coefficients of the equations.

We know how to compute 3×3 determinants, and we then find

$$x = -\tfrac{1}{5}, \qquad y = 0, \qquad z = \tfrac{2}{5}.$$

Exercise VII, §4

1. Solve the following systems of linear equations.

(a) $3x + y - z = 0$
    $x + y + z = 0$
       $y - z = 1$

(b) $2x - y + z = 1$
    $x + 3y - 2z = 0$
    $4x - 3y + z = 2$

(c) $4x + y + z + w = 1$
    $x - y + 2z - 3w = 0$
    $2x + y + 3z + 5w = 0$
    $x + y - z - w = 2$

(d) $x + 2y - 3z + 5w = 0$
    $2x + y - 4z - w = 1$
    $x + y + z + w = 0$
    $-x - y - z + w = 4$

VII, §5. Inverse of a Matrix

We consider first a special case. Let

$$A = \begin{pmatrix} a & b \\ c & d \end{pmatrix}$$

be a 2×2 matrix, and assume that its determinant $ad - bc \neq 0$. We wish to find an inverse for A, that is a 2×2 matrix

$$X = \begin{pmatrix} x & y \\ z & w \end{pmatrix}$$

such that

$$AX = XA = I.$$

Let us look at the first requirement, $AX = I$, which written out in full, looks like this:

$$\begin{pmatrix} a & b \\ c & d \end{pmatrix}\begin{pmatrix} x & y \\ z & w \end{pmatrix} = \begin{pmatrix} 1 & 0 \\ 0 & 1 \end{pmatrix}.$$

Let us look at the first column of AX. We must solve the equations

$$ax + bz = 1,$$

$$cx + dz = 0.$$

This is a system of two equations in two unknowns, x and z, which we know how to solve. Similarly, looking at the second column, we see that we must solve a system of two equations in the unknowns y, w, namely

$$ay + bw = 0,$$
$$cy + dw = 1.$$

Example. Let

$$A = \begin{pmatrix} 2 & 1 \\ 4 & 3 \end{pmatrix}.$$

We seek a matrix X such that $AX = I$. We must therefore solve the systems of linear equations

$$\begin{array}{ccc} 2x + z = 1, & & 2y + w = 0, \\ 4x + 3z = 0, & \text{and} & 4y + 3w = 1. \end{array}$$

By the ordinary method of solving two equations in two unknowns, we find

$$x = 3/2, \qquad z = -2, \qquad \text{and} \qquad y = -\tfrac{1}{2}, \qquad w = 1.$$

Thus the matrix

$$X = \begin{pmatrix} 3/2 & -\tfrac{1}{2} \\ -2 & 1 \end{pmatrix}$$

is such that $AX = I$. The reader will also verify by direct multiplication that $XA = I$. This solves for the desired inverse.

Similarly, in the 3×3 case, we would find three systems of linear equations, corresponding to the first column, the second column, and the third column. Each system could be solved to yield the inverse. We shall now give the general argument.

Let A be an $n \times n$ matrix. If B is a matrix such that $AB = I$ and $BA = I$ ($I =$ unit $n \times n$ matrix), then we called B an **inverse** of A, and we write $B = A^{-1}$. If there exists an inverse of A, then it is unique. Indeed, let C be an inverse of A. Then $CA = I$. Multiplying by B on the right, we obtain $CAB = B$. But $CAB = C(AB) = CI = C$. Hence $C = B$. A similar argument works for $AC = I$.

Theorem 5.1. *Let* $A = (a_{ij})$ *be an* $n \times n$ *matrix, and assume that* $D(A) \neq 0$. *Then* A *is invertible. Let* E^j *be the* j-th *column unit vector, and let*

$$b_{ij} = \frac{D(A^1, \dots, E^j, \dots, A^n)}{D(A)},$$

where E^j occurs in the i-th place. Then the matrix $B = (b_{ij})$ is an inverse for A.

Proof. Let $X = (x_{ij})$ be an unknown $n \times n$ matrix. We wish to solve for the components x_{ij}, so that they satisfy $AX = I$. From the definition of products of matrices, this means that for each j, we must solve

$$E^j = x_{1j}A^1 + \cdots + x_{nj}A^n.$$

This is a system of linear equations, which can be solved uniquely by Cramer's rule, and we obtain

$$x_{ij} = \frac{D(A^1,\ldots,E^j,\ldots,A^n)}{D(A)},$$

which is the formula given in the theorem.

We must still prove that $XA = I$. Note that $D({}^tA) \neq 0$. Hence by what we have already proved, we can find a matrix Y such that ${}^tAY = I$. Taking transposes, we obtain ${}^tYA = I$. Now we have

$$I = {}^tY(AX)A = {}^tYA(XA) = XA,$$

thereby proving what we want, namely that $X = B$ is an inverse for A.

We can write out the components of the matrix B in Theorem 5.1 as follows:

$$b_{ij} = \frac{\begin{vmatrix} a_{11} & \cdots & 0 & \cdots & a_{1n} \\ \vdots & & \vdots & & \vdots \\ a_{j1} & \cdots & 1 & \cdots & a_{jn} \\ \vdots & & \vdots & & \vdots \\ a_{n1} & \cdots & 0 & \cdots & a_{nn} \end{vmatrix}}{\text{Det}(A)}.$$

If we expand the determinant in the numerator according to the i-th column, then all terms but one are equal to 0, and hence we obtain the numerator of b_{ij} as a subdeterminant of $\text{Det}(A)$. Let A_{ij} be the matrix obtained from A by deleting the i-th row and the j-th column. Then

$$b_{ij} = \frac{(-1)^{i+j}\text{Det}(A_{ji})}{\text{Det}(A)}$$

(note the reversal of indices!) and thus we have the formula

$$\boxed{A^{-1} = \text{transpose of } \left(\frac{(-1)^{i+j}\,\text{Det}(A_{ij})}{\text{Det}(A)} \right).}$$

A square matrix whose determinant is $\neq 0$, or equivalently which admits an inverse, is called **non-singular**.

Example. Find the inverse of the matrix

$$A = \begin{pmatrix} 3 & 1 & -2 \\ -1 & 1 & 2 \\ 1 & -2 & 1 \end{pmatrix}.$$

By the formula, we have

$$^tA^{-1} = \left(\frac{(-1)^{i+j} \operatorname{Det}(A_{ij})}{\operatorname{Det}(A)} \right).$$

For $i = 1$, $j = 1$ the matrix A_{11} is obtained by deleting the first row and first column, that is

$$A_{11} = \begin{pmatrix} 1 & 2 \\ -2 & 1 \end{pmatrix}$$

and $\operatorname{Det}(A_{11}) = 1 - (-4) = 5$.

For $i = 1$, $j = 2$, the matrtix A_{12} is obtained by deleting the first row and second column, that is

$$A_{12} = \begin{pmatrix} -1 & 2 \\ 1 & 1 \end{pmatrix}$$

and $\operatorname{Det}(A_{12}) = -1 - 2 = -3$.

For $i = 1$, $j = 3$, the matrix A_{13} is obtained by deleting the first row and third column, that is

$$A_{13} = \begin{pmatrix} -1 & 1 \\ 1 & -2 \end{pmatrix}$$

and $\operatorname{Det}(A_{13}) = 2 - 1 = 1$.

We can compute $\operatorname{Det}(A) = 16$. Then the first row of $^tA^{-1}$ is

$$\tfrac{1}{16}(5, 3, 1).$$

Therefore the first column of A^{-1} is

$$\tfrac{1}{16}\begin{pmatrix} 5 \\ 3 \\ 1 \end{pmatrix}.$$

Observe the sign changes due to the sign pattern $(-1)^{i+j}$.

We leave the computation of the other columns of A^{-1} to the reader.

We shall assume without proof:

Theorem 5.2. *For any two $n \times n$ matrices A, B the determinant of the product is equal to the product of the determinants, that is*

$$\text{Det}(AB) = \text{Det}(A) \ \text{Det}(B).$$

Then as a special case, we find that for an invertible matrix A,

$$\boxed{\text{Det}(A^{-1}) = \text{Det}(A)^{-1}.}$$

Indeed, we have

$$AA^{-1} = I,$$

and applying the rule for a product, we get

$$D(A)D(A^{-1}) = 1,$$

thus proving the formula for the inverse.

Exercises VI, §5

1. Using determinants, find the inverses of the matrices in Chapter II, §5.
2. Write down explicitly the inverse of a 2×2 matrix

$$\begin{pmatrix} a & b \\ c & d \end{pmatrix}$$

assuming that $ad - bc \neq 0$.

VII, §6. Determinants as Area and Volume

It is remarkable that the determinant has an interpretation as a volume. We discuss first the 2-dimensional case, and thus speak of area, although we write Vol for the area of a 2-dimensional figure, to keep the terminology which generalizes to higher dimensions.

Consider the parallelogram spanned by two vectors v, w.

By definition, this parallelogram is the set of all linear combinations

$$t_1 v + t_2 w \qquad \text{with} \qquad 0 \leq t_i \leq 1.$$

Figure 1

We view v, w as column vectors, and can thus form their determinant $D(v, w)$. This determinant may be positive or negative since.

$$D(v, \ w) = -D(w, \ v).$$

Thus the determinant itself cannot be the area of this parallelogram, since area is always ≥ 0. However, we shall prove:

Theorem 6.1. *The area of the parallelogram spanned by v, w is equal to the absolute value of the determinant, namely $|D(v, w)|$.*

To prove Theorem 6.1, we introduce the notion of oriented area. Let $P(v, w)$ be the parallelogram spanned by v and w. We denote by $\mathrm{Vol}_0(v, w)$ the area of $P(v, w)$ if the determinant $D(v, w) \geq 0$, and minus the area of $P(v, w)$ if the determinant $D(v, w) < 0$. Thus at least $\mathrm{Vol}_0(v, w)$ has the same sign as the determinant, and we call $\mathrm{Vol}_0(v, w)$ the **oriented area**. We denote by $\mathrm{Vol}(v, w)$ the area of the parallelogram spanned by v, w. Hence $\mathrm{Vol}_0(v, w) = \pm\, \mathrm{Vol}(v, w)$.

To prove Theorem 6.1, it will suffice to prove:

The oriented area is equal to the determinant. In other words,

$$\mathrm{Vol}_0(v, w) = D(v, w).$$

Now to prove this, it will suffice to prove that Vol_0 satisfies the three properties characteristic of a determinant, namely:

1. Vol_0 is linear in each variable v and w.
2. $\mathrm{Vol}_0(v, v) = 0$ for all v.
3. $\mathrm{Vol}_0(E^1, E^2) = 1$ if E^1, E^2 are the standard unit vectors.

We know that these three properties characterize determinants, and this was proved in Theorem 1.1. For the convenience of the reader, we repeat the argument here very briefly. We assume that we have a function g satisfying these three properties (with g replacing Vol_0). Then for any vectors

$$v = aE^1 + cE^2 \qquad \text{and} \qquad w = bE^1 + dE^2$$

we have

$$g(aE^1 + cE^2, bE^1 + dE^2) = abg(E^1, E^1) + adg(E^1, E^2)$$
$$+ cbg(E^2, E^1) + cdg(E^2, E^2).$$

The first and fourth term are equal to 0. By Exercise 1,

$$g(E^2, E^1) = -g(E^1, E^2)$$

and hence

$$g(v, w) = (ad - bc)g(E^1, E^2) = ad - bc.$$

This proves what we wanted.

In order to prove that Vol_0 satisfies the three properties, we shall use simple properties of area (or volume) like the following: The area of a line segment is equal to 0. If A is a certain region, then the area of A is the same as the area of a translation of A, i.e. the same as the area of the region A_w (consisting of all points $v + w$ with $v \in A$). If A, B are regions which are disjoint or such that their common points have area equal to 0, then

$$\text{Vol}(A \cup B) = \text{Vol}(A) + \text{Vol}(B).$$

Consider now Vol_0. The last two properties are obvious. Indeed, the parallelogram spanned by v, v is simply a line segment, and its 2-dimensional area is therefore equal to 0. Thus property **2** is satisfied. As for the third property, the parallelogram spanned by the unit vectors E^1, E^2 is simply the unit square, whose area is 1. Hence in this case we have

$$\text{Vol}_0(E^1, E^2) = 1.$$

The harder property is the first. The reader who has not already done so, should now read the geometric applications of Chapter III, §2 before reading the rest of this proof, which we shall base on geometric considerations concerning area.

We shall need a lemma.

Lemma 6.2. *If v, w are linearly dependent, then $\text{Vol}_0(v, w) = 0$.*

Proof. Suppose that we can write

$$av + bw = 0$$

Figure 2

with a or $b \neq 0$. Say $a \neq 0$. Then

$$v = -\frac{b}{a} \, w = cw$$

so that v, w lie on the same straight line, and the parallelogram spanned by v, w is a line segment (Fig. 2). Hence $\mathrm{Vol}_0(v, w) = 0$, thus proving the lemma.

We also know that when v, w are linearly dependent, then $D(v, w) = 0$, so in this trivial case, our theorem is proved. In the subsequent lemmas, we assume that v, w are linearly independent.

Lemma 6.3. *Assume that v, w are linearly independent, and let n be a positive integer. Then*

$$\mathrm{Vol}(nv, \, w) = n \, \mathrm{Vol}(v, \, w).$$

Proof. The parallelogram spanned by nv and w consists of n parallelograms as shown in the following picture.

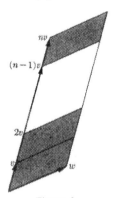

Figure 3

These n parallelograms are simply the translations of $P(v, w)$ by $v, 2v, \ldots, (n-1)v$, and each translation of $P(v, w)$ has the same area as $P(v, w)$. These translations have only line segments in common, and hence

$$\mathrm{Vol}(nv, \, w) = n \, \mathrm{Vol}(v, \, w)$$

as desired.

Corollary 6.4. *Assume that v, w are linearly independent and let n be a positive integer. Then*

$$\text{Vol}\left(\frac{1}{n}\,v,\ w\right) = \frac{1}{n}\,\text{Vol}(v, w).$$

If m, n are positive integers, then

$$\text{Vol}\left(\frac{m}{n}\,v,\ w\right) = \frac{m}{n}\,\text{Vol}(v, w).$$

Proof. Let $v_1 = (1/n)v$. By the lemma, we know that

$$\text{Vol}(nv_1, w) = n\,\text{Vol}(v_1, w).$$

This is merely a reformulation of our first assertion, since $nv_1 = v$. As for the second assertion, we write $m/n = m \cdot 1/n$ and apply the proved statements successively:

$$\text{Vol}\left(m \cdot \frac{1}{n}v, w\right) = m\,\text{Vol}\left(\frac{1}{n}v, w\right)$$

$$= m \cdot \frac{1}{n}\,\text{Vol}(v, w)$$

$$= \frac{m}{n}\,\text{Vol}(v, w).$$

Lemma 6.5. $\text{Vol}(-v, w) = \text{Vol}(v, w)$.

Proof. The parallelogram spanned by $-v$ and w is a translation by $-v$ of the parallelogram $P(v, w)$. Hence $P(v, w)$ and $P(-v, w)$ have the same area. (Cf. Fig. 4.)

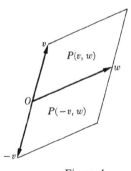

Figure 4

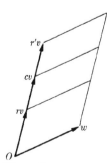

Figure 5

Lemma 6.6. *If c is any real number > 0, then*

$$\text{Vol}(cv, w) = c \ \text{Vol}(v, w).$$

Proof. Let r, r' be rational numbers such that $0 < r < c < r'$ (Fig. 5). Then

$$P(rv, w) \subset P(cv, w) \subset P(r'v, w).$$

Hence by Lemma 6.3,

$$\begin{aligned}
r \ \text{Vol}(v, w) &= \text{Vol}(rv, w) \\
&\leq \text{Vol}(cv, w) \\
&\leq \text{Vol}(r'v, w) \\
&= r' \ \text{Vol}(v, w).
\end{aligned}$$

Letting r and r' approach c as a limit, we find that

$$\text{Vol}(cv, w) = c \ \text{Vol}(v, w),$$

as was to be shown.

From Lemmas 6.5 and 6.6 we can now prove that

$$\boxed{\text{Vol}_0(cv, w) = c \ \text{Vol}_0(v, w)}$$

for any real number c, and any vectors v, w. Indeed, if v, w are linearly dependent, then both sides are equal to 0. If v, w are linearly independent, we use the definition of Vol_0 and Lemmas 6.5, 6.6. Say $D(v, w) > 0$ and c is negative, $c = -d$. Then $D(cv, w) \leq 0$ and consequently

$$\begin{aligned}
\text{Vol}_0(cv, w) = -\text{Vol}(cv, w) &= -\text{Vol}(-dv, w) \\
&= -\text{Vol}(dv, w) \\
&= -d \ \text{Vol}(v, w) \\
&= c \ \text{Vol}(v, w) = c \ \text{Vol}_0(v, w).
\end{aligned}$$

A similar argument works when $D(v, w) \leq 0$. We have therefore proved one of the conditions of linearity of the function Vol_0. The analogous property of course works on the other side, namely

$$\boxed{\text{Vol}_0(v, cw) = c \ \text{Vol}_0(v, w).}$$

For the other condition, we again have a lemma.

Lemma 6.7. *Assume that v, w are linearly independent. Then*

$$\text{Vol}(v + w, \ w) = \text{Vol}(v, \ w).$$

Proof. We have to prove that the parallelogram spanned by v, w has the same area as the parallelogram spanned by $v + w$, w.

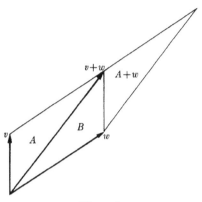

Figure 6

The parallelogram spanned by v, w consists of two triangles A and B as shown in the picture. The parallelogram spanned by $v + w$ and w consists of the triangles B and the translation of A by w. Since A and $A + w$ have the same area, we get:

$$\text{Vol}(v, \ w) = \text{Vol}(A) + \text{Vol}(B) = \text{Vol}(A + w) + \text{Vol}(B) = \text{Vol}(v + w, \ w),$$

as was to be shown.

We are now in a position to deal with the second property of linearity. Let w be a fixed non-zero vector in the plane, and let v be a vector such that $\{v, w\}$ is a basis of the plane. We shall prove that for any numbers c, d we have

(1) $$\text{Vol}_0(cv + dw, \ w) = c \ \text{Vol}_0(v, \ w).$$

Indeed, if $d = 0$, this is nothing but what we have shown previously. If $d \neq 0$, then again by what has been shown previously,

$$d \ \text{Vol}_0(cv + dw, \ w) = \text{Vol}_0(cv + dw, \ dw) = c \ \text{Vol}_0(v, \ dw) = cd \ \text{Vol}_0(v, \ w).$$

Canceling d yields relation (1).

From this last formula, the linearity now follows. Indeed, if

$$v_1 = c_1 v + d_1 w \qquad \text{and} \qquad v_2 = c_2 v + d_2 w,$$

then

$$\begin{aligned}
\text{Vol}_0(v_1 + v_2, w) &= \text{Vol}_0((c_1 + c_2)v + (d_1 + d_2)w, w) \\
&= (c_1 + c_2)\,\text{Vol}_0(v, w) \\
&= c_1 \text{Vol}_0(v, w) + c_2\,\text{Vol}_0(v, w) \\
&= \text{Vol}_0(v_1, w) + \text{Vol}_0(v_2, w).
\end{aligned}$$

This concludes the proof of the fact that

$$\text{Vol}_0(v, \ w) = D(v, \ w),$$

and hence of Theorem 6.1.

Remark 1. The proof given above is slightly long, but each step is quite simple. Furthermore, when one wishes to generalize the proof to higher dimensional space (even 3-space), one can give an entirely similar proof. The reason for this is that the conditions characterizing a determinant involve only two coordinates at a time and thus always take place in some 2-dimensional plane. Keeping all but two coordinates fixed, the above proof then can be extended at once. Thus for instance in 3-space, let us denote by $P(u, v, w)$ the box spanned by vectors u, v, w (Fig. 7), namely all combinations

$$t_1 u + t_2 v + t_3 w \qquad \text{with} \qquad 0 \leq t_i \leq 1.$$

Let $\text{Vol}(u, v, w)$ be the volume of this box.

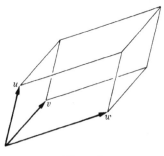

Figure 7

Theorem 6.8. *The volume of the box spanned by u, v, w is the absolute value of the determinant* $|D(u, v, w)|$. *That is,*

$$\text{Vol}(u, v, w) = |D(u, v, w)|.$$

The proof follows exactly the same pattern as in the two-dimensional case. Indeed, the volume of the cube spanned by the unit vectors is 1. If two of the vectors u, v, w are equal, then the box is actually a 2-dimensional parallelogram, whose 3-dimensional volume is 0. Finally, the proof of linearity is the same, because all the action took place either in one or in two variables. The other variables can just be carried on in the notation but they did not enter in an essential way in the proof.

Similarly, one can define n-dimensional volumes, and the corresponding theorem runs as follows.

Theorem 6.9. *Let* $v_1, \ldots, v_n$ *be elements of* $\mathbf{R}^n$. *Let* $\text{Vol}(v_1, \ldots, v_n)$ *be the n-dimensional volume of the n-dimensional box spanned by* $v_1, \ldots, v_n$. *Then*

$$\text{Vol}(v_1, \ldots, v_n) = |D(v_1, \ldots, v_n)|.$$

Of course, the n-dimensional box spanned by $v_1, \ldots, v_n$ is the set of linear combinations

$$\sum_{i=1}^{n} t_i v_i \qquad \text{with} \qquad 0 \leq t_i \leq 1.$$

Remark 2. We have used geometric properties of area to carry out the above proof. One can lay foundations for all this purely analytically. If the reader is interested, cf. my book *Undergraduate Analysis*.

Remark 3. In the special case of dimension 2, one could actually have given a simpler proof that the determinant is equal to the area. But we chose to give the slightly more complicated proof because it is the one which generalizes to the 3-dimensional, or n-dimensional case.

We interpret Theorem 6.1 in terms of linear maps. Given vectors v, w in the plane, we know that there exists a unique linear map.

$$L: \mathbf{R}^2 \rightarrow \mathbf{R}^2$$

such that $L(E^1) = v$ and $L(E^2) = w$. In fact, if

$$v = aE^1 + cE^2, \qquad w = bE^1 + dE^2,$$

then the matrix associated with the linear map is

$$\begin{pmatrix} a & b \\ c & d \end{pmatrix}.$$

Furthermore, if we denote by C the unit square spanned by E^1, E^2, and by P the parallelogram spanned by v, w, then P is the image under L of C, that is $L(C) = P$. Indeed, as we have seen, for $0 \leqq t_i \leqq 1$ we have

$$L(t_1 E^1 + t_2 E^2) = t_1 L(E^1) + t_2 L(E^2) = t_1 v + t_2 w.$$

If we define the determinant of a linear map to be the determinant of its associated matrix, we conclude that

$(*)$ (Area of P) $= |\text{Det}(L)|$.

To take a numerical example, the area of the parallelogram spanned by the vectors $(2, 1)$ and $(3, -1)$ (Fig. 8) is equal to the absolute value of

$$\begin{vmatrix} 2 & 1 \\ 3 & -1 \end{vmatrix} = -5$$

and hence is equal to 5.

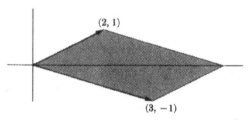

Figure 8

Theorem 6.10. *Let P be a parallelogram spanned by two vectors. Let $L: \mathbf{R}^2 \to \mathbf{R}^2$ be a linear map. Then*

Area of $L(P) = |\text{Det } L|$ (Area of P).

Proof. Suppose that P is spanned by two vectors v, w. Then $L(P)$ is spanned by $L(v)$ and $L(w)$. (Cf. Fig. 9.) There is a linear map $L_1: \mathbf{R}^2 \to \mathbf{R}^2$ such that

$$L_1(E^1) = v \quad \text{and} \quad L_1(E^2) = w.$$

Then $P = L_1(C)$, where C is the unit square, and

$$L(P) = L(L_1(C)) = (L \circ L_1)(C).$$

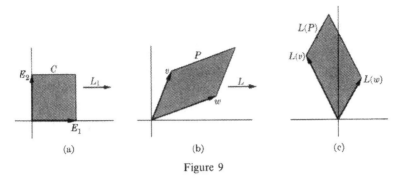

Figure 9

By what we proved above in (∗), we obtain

$$\text{Vol } L(P) = |\text{Det}(L \circ L_1)| = |\text{Det}(L)\,\text{Det}(L_1)| = |\text{Det}(L)|\text{Vol}(P),$$

thus proving our assertion.

Corollary 6.11. *For any rectangle R with sides parallel to the axes, and any linear map $L: \mathbf{R}^2 \to \mathbf{R}^2$ we have*

$$\text{Vol } L(R) = |\text{Det}(L)|\ \text{Vol}(R).$$

Proof. Let c_1, c_2 be the lengths of the sides of R. Let R_1 be the rectangle spanned by $c_1 E^1$ and $c_2 E^2$. Then R is the translation of R_1 by some vector, say $R = R_1 + u$. Then

$$L(R) = L(R_1 + u) = L(R_1) + L(u)$$

is the translation of $L(R_1)$ by $L(u)$. (Cf. Fig. 10.) Since area does not change under translation, we need only apply Theorem 6.1 to conclude the proof.

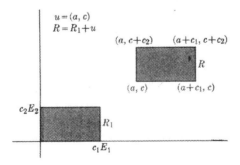

Figure 10

Exercises VII, §6

1. If $g(v, w)$ satisfies the first two axioms of a determinant, prove that

$$g(v, w) = -g(w, v)$$

for all vectors v, w. This fact was used in the uniqueness proof. [*Hint:* Expand $g(v + w, v + w) = 0$.]

2. Find the area of the parallelogram spanned by the following vectors.
   (a) (2, 1) and (-4, 5)   (b) (3, 4) and (-2, -3)

3. Find the area of the parallelogram such that three corners of the parallelogram are given by the following points.
   (a) (1, 1), (2, -1), (4, 6)   (b) (-3, 2), (1, 4), (-2, -7)
   (c) (2, 5), (-1, 4), (1, 2)   (d) (1, 1), (1, 0), (2, 3)

4. Find the volume of the parallelepiped spanned by the following vectors in 3-space.
   (a) (1, 1, 3), (1, 2, -1), (1, 4, 1)   (b) (1, -1, 4), (1, 1, 0), (-1, 2, 5)
   (c) (-1, 2, 1), (2, 0, 1), (1, 3, 0)   (d) (-2, 2, 1), (0, 1, 0), (-4, 3, 2)

CHAPTER VIII

Eigenvectors and Eigenvalues

This chapter gives the basic elementary properties of eigenvectors and eigenvalues. We get an application of determinants in computing the characteristic polynomial. In §3, we also get an elegant mixture of calculus and linear algebra by relating eigenvectors with the problem of finding the maximum and minimum of a quadratic function on the sphere. Most students taking linear algebra will have had some calculus, but the proof using complex numbers instead of the maximum principle can be used to get real eigenvalues of a symmetric matrix if the calculus has to be avoided. Basic properties of the complex numbers will be recalled in an appendix.

VIII, §1. Eigenvectors and Eigenvalues

Let V be a vector space and let

$$A: V \to V$$

be a linear map of V into itself. An element $v \in V$ is called an **eigenvector** of A if there exists a number λ such that $Av = \lambda v$. *If $v \neq O$ then λ is uniquely determined*, because $\lambda_1 v = \lambda_2 v$ implies $\lambda_1 = \lambda_2$. In this case, we say that λ is an **eigenvalue** of A belonging to the eigenvector v. We also say that v is an eigenvector with the eigenvalue λ. Instead of eigenvector and eigenvalue, one also uses the terms **characteristic vector** and **characteristic value**.

If A is a square $n \times n$ matrix then an **eigenvector** of A is by definition an eigenvector of the linear map of $\mathbf{R}^n$ into itself represented by this matrix. Thus an eigenvector X of A is a (column) vector of $\mathbf{R}^n$ for which there exists $\lambda \in \mathbf{R}$ such that $AX = \lambda X$.

Example 1. Let V be the vector space over $\mathbf{R}$ consisting of all infinitely differentiable functions. Let $\lambda \in \mathbf{R}$. Then the function f such that $f(t) = e^{\lambda t}$ is an eigenvector of the derivative d/dt because $df/dt = \lambda e^{\lambda t}$.

Example 2. Let

$$A = \begin{pmatrix} a_1 & \cdots & 0 \\ \vdots & \ddots & \vdots \\ 0 & \cdots & a_n \end{pmatrix}$$

be a diagonal matrix. Then every unit vector E^i $(i = 1, \ldots, n)$ is an eigenvector of A. In fact, we have $AE^i = a_i E^i$:

$$\begin{pmatrix} a_1 & 0 & \cdots & 0 \\ 0 & a_2 & \cdots & 0 \\ \vdots & \vdots & & \vdots \\ 0 & 0 & \cdots & a_n \end{pmatrix} \begin{pmatrix} 0 \\ \vdots \\ 1 \\ \vdots \\ 0 \end{pmatrix} = \begin{pmatrix} 0 \\ \vdots \\ a_i \\ \vdots \\ 0 \end{pmatrix}.$$

Example 3. If $A: V \to V$ is a linear map, and v is an eigenvector of A, then for any non-zero scalar c, cv is also an eigenvector of A, with the same eigenvalue.

Theorem 1.1. *Let V be a vector space and let $A: V \to V$ be a linear map. Let $\lambda \in \mathbf{R}$. Let V_λ be the subspace of V generated by all eigenvectors of A having λ as eigenvalue. Then every non-zero element of V_λ is an eigenvector of A having λ as eigenvalue.*

Proof. Let $v_1, v_2 \in V$ be such that $Av_1 = \lambda v_1$ and $Av_2 = \lambda v_2$. Then

$$A(v_1 + v_2) = Av_1 + Av_2 = \lambda v_1 + \lambda v_2 = \lambda(v_1 + v_2).$$

If $c \in K$ then $A(cv_1) = cAv_1 = c\lambda v_1 = \lambda cv_1$. This proves our theorem.

The subspace V_λ in Theorem 1.1 is called the **eigenspace** of A belonging to λ.

Note. If v_1, v_2 are eigenvectors of A with different eigenvalues $\lambda_1 \neq \lambda_2$ then of course $v_1 + v_2$ is *not* an eigenvector of A. In fact, we have the following theorem:

Theorem 1.2. *Let V be a vector space and let $A: V \to V$ be a linear map.*

Let $v_1, \ldots, v_m$ be eigenvectors of A, with eigenvalues $\lambda_1, \ldots, \lambda_m$ respectively. Assume that these eigenvalues are distinct, i.e.

$$\lambda_i \neq \lambda_j \qquad if \qquad i \neq j.$$

Then $v_1, \ldots, v_m$ are linearly independent.

Proof. By induction on m. For $m = 1$, an element $v_1 \in V$, $v_1 \neq O$ is linearly independent. Assume $m > 1$. Suppose that we have a relation

(∗) $$c_1 v_1 + \cdots + c_m v_m = O$$

with scalars c_i. We must prove all $c_i = 0$. We multiply our relation (∗) by λ_1 to obtain

$$c_1 \lambda_1 v_1 + \cdots + c_m \lambda_1 v_m = O.$$

We also apply A to our relation (∗). By linearity, we obtain

$$c_1 \lambda_1 v_1 + \cdots + c_m \lambda_m v_m = O.$$

We now subtract these last two expressions, and obtain

$$c_2(\lambda_2 - \lambda_1)v_2 + \cdots + c_m(\lambda_m - \lambda_1)v_m = O.$$

Since $\lambda_j - \lambda_1 \neq 0$ for $j = 2, \ldots, m$ we conclude by induction that $c_2 = \cdots = c_m = 0$. Going back to our original relation, we see that $c_1 v_1 = O$, whence $c_1 = 0$, and our theorem is proved.

Example 4. Let V be the vector space consisting of all differentiable functions of a real variable t. Let $\alpha_1, \ldots, \alpha_m$ be distinct numbers. The functions

$$e^{\alpha_1 t}, \ldots, e^{\alpha_m t}$$

are eigenvectors of the derivative, with distinct eigenvalues $\alpha_1, \ldots, \alpha_m$, and hence are linearly independent.

Remark 1. In Theorem 1.2, suppose V is a vector space of dimension n and $A: V \to V$ is a linear map having n eigenvectors $v_1, \ldots, v_n$ whose eigenvalues $\lambda_1, \ldots, \lambda_n$ are distinct. Then $\{v_1, \ldots, v_n\}$ is a basis of V.

Remark 2. One meets a situation like that of Theorem 1.2 in the theory of linear differential equations. Let $A = (a_{ij})$ be an $n \times n$ matrix, and let

$$F(t) = \begin{pmatrix} f_1(t) \\ \vdots \\ f_n(t) \end{pmatrix}$$

be a column vector of functions satisfying the equation

$$\frac{dF}{dt} = AF(t).$$

In terms of the coordinates, this means that

$$\frac{df_i}{dt} = \sum_{j=1}^{n} a_{ij} f_j(t).$$

Now suppose that A is a diagonal matrix,

$$A = \begin{pmatrix} a_1 & 0 & \cdots & 0 \\ \vdots & \vdots & & \vdots \\ 0 & 0 & \cdots & a_n \end{pmatrix} \quad \text{with } a_i \neq 0 \quad \text{all } i.$$

Then each function $f_i(t)$ satisfies the equation

$$\frac{df_i}{dt} = a_i f_i(t).$$

By calculus, there exist numbers $c_1,\ldots,c_n$ such that for $i = 1,\ldots,n$ we have

$$f_i(t) = c_i e^{a_i t}.$$

[Proof: if $df/dt = af(t)$, then the derivative of $f(t)/e^{at}$ is 0, so $f(t)/e^{at}$ is constant.] Conversely, if $c_1,\ldots,c_n$ are numbers, and we let

$$F(t) = \begin{pmatrix} c_1 e^{a_1 t} \\ \vdots \\ c_n e^{a_n t} \end{pmatrix}.$$

Then $F(t)$ satisfies the differential equation

$$\frac{dF}{dt} = AF(t).$$

Let V be the set of solutions $F(t)$ for the differential equation $dF/dt = AF(t)$. Then V is immediately verified to be a vector space, and the above argument shows that the n elements

$$\begin{pmatrix} e^{a_1 t} \\ \vdots \\ 0 \\ 0 \end{pmatrix}, \begin{pmatrix} 0 \\ e^{a_2 t} \\ \vdots \\ 0 \end{pmatrix}, \quad \ldots, \quad \begin{pmatrix} 0 \\ 0 \\ \vdots \\ e^{a_n t} \end{pmatrix}$$

form a basis for V. Furthermore, these elements are eigenvectors of A, and also of the derivative (viewed as a linear map).

The above is valid if A is a diagonal matrix. If A is not diagonal, then we try to find a basis such that we can represent the linear map A by a diagonal matrix. We do not go into this type of consideration here.

Exercises VIII, §1

Let a be a number $\neq 0$.

1. Prove that the eigenvectors of the matrix

$$\begin{pmatrix} 1 & a \\ 0 & 1 \end{pmatrix}$$

generate a 1-dimensional space, and give a basis for this space.

2. Prove that the eigenvectors of the matrix

$$\begin{pmatrix} 2 & 0 \\ 0 & 2 \end{pmatrix}$$

generate a 2-dimensional space and give a basis for this space. What are the eigenvalues of this matrix?

3. Let A be a diagonal matrix with diagonal elements $a_{11}, \dots, a_{nn}$. What is the dimension of the space generated by the eigenvectors of A? Exhibit a basis for this space, and give the eigenvalues.

4. Show that if $\theta \in \mathbf{R}$, then the matrix

$$A = \begin{pmatrix} \cos\theta & \sin\theta \\ \sin\theta & -\cos\theta \end{pmatrix}$$

always has an eigenvector in $\mathbf{R}^2$, and in fact that there exists a vector v_1 such that $Av_1 = v_1$. [*Hint:* Let the first component of v_1 be

$$x = \frac{\sin\theta}{1 - \cos\theta}$$

if $\cos\theta \neq 1$. Then solve for y. What if $\cos\theta = 1$?]

5. In Exercise 4, let v_2 be a vector of $\mathbf{R}^2$ perpendicular to the vector v_1 found in that exercise. Show that $Av_2 = -v_2$. Define this to mean that A is a reflection.

6. Let

$$R(\theta) = \begin{pmatrix} \cos\theta & -\sin\theta \\ \sin\theta & \cos\theta \end{pmatrix}$$

be the matrix of a rotation. Show that $R(\theta)$ does not have any real eigen-values unless $R(\theta) = \pm I$. [It will be easier to do this exercise after you have read the next section.]

7. Let V be a finite dimensional vector space. Let A, B be linear maps of V into itself. Assume that $AB = BA$. Show that if v is an eigenvector of A, with eigenvalue λ, then Bv is an eigenvector of A, with eigenvalue λ also if $Bv \neq O$.

VIII, §2. The Characteristic Polynomial

We shall now see how we can use determinants to find the eigenvalue of a matrix.

Theorem 2.1. *Let V be a finite dimensional vector space, and let λ be a number. Let $A: V \to V$ be a linear map. Then λ is an eigenvalue of A if and only if $A - \lambda I$ is not invertible.*

Proof. Assume that λ is an eigenvalue of A. Then there exists an element $v \in V$, $v \neq O$ such that $Av = \lambda v$. Hence $Av - \lambda v = O$, and $(A - \lambda I)v = O$. Hence $A - \lambda I$ has a non-zero kernel, and $A - \lambda I$ cannot be invertible. Conversely, assume that $A - \lambda I$ is not invertible. By Theorem 2.4 of Chapter 5, we see that $A - \lambda I$ must have a non-zero kernel, meaning that there exists an element $v \in V$, $v \neq O$ such that $(A - \lambda I)v = O$. Hence $Av - \lambda v = O$, and $Av = \lambda v$. Thus λ is an eigen-value of A. This proves our theorem.

Let A be an $n \times n$ matrix, $A = (a_{ij})$. We define the **characteristic polynomial** P_A of A to be the determinant

$$P_A(t) = \text{Det}(tI - A),$$

or written out in full,

$$P(t) = \begin{vmatrix} t - a_{11} & -a_{12} & \cdots & -a_{1n} \\ \vdots & \vdots & & \vdots \\ -a_{n1} & -a_{n2} & \cdots & t - a_{nn} \end{vmatrix} = \begin{vmatrix} t - a_{11} & & & \\ & \ddots & & -a_{ij} \\ & -a_{ij} & & \\ & & & t - a_{nn} \end{vmatrix}.$$

We can also view A as a linear map from $\mathbf{R}^n$ to $\mathbf{R}^n$, and we also say that $P_A(t)$ is the **characteristic polynomial of this linear map**.

Example 1. The characteristic polynomial of the matrix

$$A = \begin{pmatrix} 1 & -1 & 3 \\ -2 & 1 & 1 \\ 0 & 1 & -1 \end{pmatrix}$$

is

$$\begin{vmatrix} t-1 & 1 & -3 \\ 2 & t-1 & -1 \\ 0 & -1 & t+1 \end{vmatrix},$$

which we expand according to the first column, to find

$$P_A(t) = t^3 - t^2 - 4t + 6.$$

For an arbitrary matrix $A = (a_{ij})$, the characteristic polynomial can be found by expanding according to the first column, and will always consist of a sum

$$(t - a_{11}) \cdots (t - a_{nn}) + \cdots.$$

Each term other than the one we have written down will have degree $< n$. Hence the characteristic polynomial is of type

$$P_A(t) = t^n + \text{terms of lower degree}.$$

Theorem 2.2. *Let A be an $n \times n$ matrix. A number λ is an eigenvalue of A if and only if λ is a root of the characteristic polynomial of A.*

Proof. Assume that λ is an eigenvalue of A. Then $\lambda I - A$ is not invertible by Theorem 2.1, and hence $\text{Det}(\lambda I - A) = 0$, by Theorem 3.1 of Chapter VII and Theorem 2.5 of Chapter V. Consequently λ is a root of the characteristic polynomial. Conversely, if λ is a root of the characteristic polynomial, then

$$\text{Det}(\lambda I - A) = 0,$$

and hence by the same Theorem 3.1 of Chapter VII we conclude that $\lambda I - A$ is not invertible. Hence λ is an eigenvalue of A by Theorem 2.1.

Theorem 2.2 gives us an explicit way of determining the eigenvalues of a matrix, *provided* that we can determine explicitly the roots of its characteristic polynomial. This is sometimes easy, especially in exercises at the end of chapters when the matrices are adjusted in such a way that one can determine the roots by inspection, or simple devices. It is considerably harder in other cases.

For instance, to determine the roots of the polynomial in Example 1, one would have to develop the theory of cubic polynomials. This can be done, but it involves formulas which are somewhat harder than the formula needed to solve a quadratic equation. One can also find methods to determine roots approximately. In any case, the determination of such methods belongs to another range of ideas than that studied in the present chapter.

Example 2. Find the eigenvalues and a basis for the eigenspaces of the matrix

$$\begin{pmatrix} 1 & 4 \\ 2 & 3 \end{pmatrix}.$$

The characteristic polynomial is the determinant

$$\begin{vmatrix} t-1 & -4 \\ -2 & t-3 \end{vmatrix} = (t-1)(t-3) - 8 = t^2 - 4t - 5 = (t-5)(t+1).$$

Hence the eigenvalues are $5, -1$.

For any eigenvalue λ, a corresponding eigenvector is a vector $\begin{pmatrix} x \\ y \end{pmatrix}$ such that

$$x + 4y = \lambda x,$$
$$2x + 3y = \lambda y,$$

or equivalently

$$(1 - \lambda)x + 4y = 0,$$
$$2x + (3 - \lambda)y = 0.$$

We give x some value, say $x = 1$, and solve for y from either equation, for instance the second to get $y = -2/(3 - \lambda)$. This gives us the eigenvector

$$X(\lambda) = \begin{pmatrix} 1 \\ -2/(3 - \lambda) \end{pmatrix}.$$

Substituting $\lambda = 5$ and $\lambda = -1$ gives us the two eigenvectors

$$X^1 = \begin{pmatrix} 1 \\ 1 \end{pmatrix} \text{ for } \lambda = 5, \quad \text{and} \quad X^2 = \begin{pmatrix} 1 \\ -\frac{1}{2} \end{pmatrix} \text{ for } \lambda = -1.$$

The eigenspace for 5 has basis X^1 and the eigenspace for -1 has basis X^2. Note that any non-zero scalar multiples of these vectors would also be bases. For instance, instead of X^2 we could take

$$\begin{pmatrix} 2 \\ -1 \end{pmatrix}.$$

Example 3. Find the eigenvalues and a basis for the eigenspaces of the matrix

$$\begin{pmatrix} 2 & 1 & 0 \\ 0 & 1 & -1 \\ 0 & 2 & 4 \end{pmatrix}.$$

The characteristic polynomial is the determinant

$$\begin{vmatrix} t-2 & -1 & 0 \\ 0 & t-1 & 1 \\ 0 & -2 & t-4 \end{vmatrix} = (t-2)^2(t-3).$$

Hence the eigenvalues are 2 and 3.

For the eigenvectors, we must solve the equations

$$(2 - \lambda)x + y = 0,$$
$$(1 - \lambda)y - z = 0,$$
$$2y + (4 - \lambda)z = 0.$$

Note the coefficient $(2 - \lambda)$ of x.

Suppose we want to find the eigenspace with eigenvalue $\lambda = 2$. Then the first equation becomes $y = 0$, whence $z = 0$ from the second equation. We can give x any value, say $x = 1$. Then the vector

$$X^1 = \begin{pmatrix} 1 \\ 0 \\ 0 \end{pmatrix}$$

is a basis for the eigenspace with eigenvalue 2.

Now suppose $\lambda \neq 2$, so $\lambda = 3$. If we put $x = 1$ then we can solve for y from the first equation to give $y = 1$, and then we can solve for z in the second equation, to get $z = -2$. Hence

$$X^2 = \begin{pmatrix} 1 \\ 1 \\ -2 \end{pmatrix}$$

is a basis for the eigenvectors with eigenvalue 3. Any non-zero scalar multiple of X^2 would also be a basis.

Example 4. The characteristic polynomial of the matrix

$$\begin{pmatrix} 1 & 1 & 2 \\ 0 & 5 & -1 \\ 0 & 0 & 7 \end{pmatrix}$$

is $(t - 1)(t - 5)(t - 7)$. Can you generalize this?

Example 5. Find the eigenvalues and a basis for the eigenspaces of the matrix in Example 4.

The eigenvalues are 1, 5, and 7. Let X be a non-zero eigenvector, say

$$X = \begin{pmatrix} x \\ y \\ z \end{pmatrix} \quad \text{also written} \quad {}^tX = (x, y, z).$$

Then by definition of an eigenvector, there is a number λ such that $AX = \lambda X$, which means

$$x + y + 2z = \lambda x,$$
$$5y - z = \lambda y,$$
$$7z = \lambda z.$$

Case 1. $z = 0$, $y = 0$. Since we want a non-zero eigenvector we must then have $x \neq 0$, in which case $\lambda = 1$ by the first equation. Let $X^1 = E^1$ be the first unit vector, or any non-zero scalar multiple to get an eigenvector with eigenvalue 1.

Case 2. $z = 0$, $y \neq 0$. By the second equation, we must have $\lambda = 5$. Give y a specific value, say $y = 1$. Then solve the first equation for x, namely

$$x + 1 = 5x, \quad \text{which gives} \quad x = \tfrac{1}{4}.$$

Let

$$\dot{X}^2 = \begin{pmatrix} \tfrac{1}{4} \\ 1 \\ 0 \end{pmatrix}$$

Then X^2 is an eigenvector with eigenvalue 5.

Case 3. $z \neq 0$. Then from the third equation, we must have $\lambda = 7$. Fix some non-zero value of z, say $z = 1$. Then we are reduced to solving the two simultaneous equations

$$x + y + 2 = 7x,$$
$$5y - 1 = 7y.$$

This yields $y = -\tfrac{1}{2}$ and $x = \tfrac{1}{4}$. Let

$$X^3 = \begin{pmatrix} \tfrac{1}{4} \\ -\tfrac{1}{2} \\ 1 \end{pmatrix}.$$

Then X^3 is an eigenvector with eigenvalue 7.

Scalar multiples of X^1, X^2, X^3 will yield eigenvectors with the same eigenvalues as X^1, X^2, X^3 respectively. Since these three vectors have distinct eigenvalues, they are linearly independent, and so form a basis of $\mathbf{R}^3$. By Exercise 14, there are no other eigenvectors.

Finally we point out that the linear algebra of matrices could have been carried out with complex coefficients. The same goes for determinants. All that is needed about numbers is that one can add, multiply, and divide by non-zero numbers, and these operations are valid with complex numbers. Then a matrix $A = (a_{ij})$ of complex numbers has eigenvalues and eigenvectors whose components are complex numbers. This is useful because of the following fundamental fact:

Every non-constant polynomial with complex coefficients has a complex root.

If A is a complex $n \times n$ matrix, then the characteristic polynomial of A has complex coefficients, and has degree $n \geq 1$, so has a complex root which is an eigenvalue. Thus over the complex numbers, a square matrix always has an eigenvalue, and a non-zero eigenvector. This is not always true over the real numbers. (Example?) In the next section, we shall see an important case when a real matrix always has a real eigenvalue.

We now give examples of computations using complex numbers for the eigenvalues and eigenvectors, even though the matrix itself has real components. It should be remembered that in the case of complex eigenvalues, the vector space is over the complex numbers, so it consists of linear combinations of the given basis elements with complex coefficients.

Example 6. Find the eigenvalues and a basis for the eigenspaces of the matrix

$$A = \begin{pmatrix} 2 & -1 \\ 3 & 1 \end{pmatrix}.$$

The characteristic polynomial is the determinant

$$\begin{vmatrix} t-2 & 1 \\ -3 & t-1 \end{vmatrix} = (t-2)(t-1) + 3 = t^2 - 3t + 5.$$

Hence the eigenvalues are

$$\frac{3 \pm \sqrt{9-20}}{2}.$$

Thus there are two distinct eigenvalues (but no real eigenvalue):

$$\lambda_1 = \frac{3 + \sqrt{-11}}{2} \quad \text{and} \quad \lambda_2 = \frac{3 - \sqrt{-11}}{2}.$$

Let $X = \begin{pmatrix} x \\ y \end{pmatrix}$ with not both x, y equal to 0. Then X is an eigenvector if and only if $AX = \lambda X$, that is:

$$2x - y = \lambda x,$$
$$3x + y = \lambda y,$$

where λ is an eigenvalue. This system is equivalent with

$$(2 - \lambda)x - y = 0,$$
$$3x + (1 - \lambda)y = 0.$$

We give x, say, an arbitrary value, for instance $x = 1$ and solve for y, so $y = (2 - \lambda)$ from the first equation. Then we obtain the eigenvectors

$$X(\lambda_1) = \begin{pmatrix} 1 \\ 2 - \lambda_1 \end{pmatrix} \quad \text{and} \quad X(\lambda_2) = \begin{pmatrix} 1 \\ 2 - \lambda_2 \end{pmatrix}.$$

Remark. We solved for y from one of the equations. This is consistent with the other because λ is an eigenvalue. Indeed, if you substitute $x = 1$ and $y = 2 - \lambda$ on the left in the second equation, you get

$$3 + (1 - \lambda)(2 - \lambda) = 0$$

because λ is a root of the characteristic polynomial.

Then $X(\lambda_1)$ is a basis for the one-dimensional eigenspace of λ_1, and $X(\lambda_2)$ is a basis for the one-dimensional eigenspace of λ_2.

Example 7. Find the eigenvalues and a basis for the eigenspaces of the matrix

$$A = \begin{pmatrix} 1 & 1 & -1 \\ 0 & 1 & 0 \\ 1 & 0 & 1 \end{pmatrix}.$$

We compute the characteristic polynomial, which is the determinant

$$\begin{vmatrix} t - 1 & -1 & 1 \\ 0 & t - 1 & 0 \\ -1 & 0 & t - 1 \end{vmatrix}$$

easily computed to be

$$P(t) = (t - 1)(t^2 - 2t + 2).$$

Now we meet the problem of finding the roots of $P(t)$ as real numbers or complex numbers. By the quadratic formula, the roots of $t^2 - 2t + 2$ are given by

$$\frac{2 \pm \sqrt{4 - 8}}{2} = 1 \pm \sqrt{-1}.$$

The whole theory of linear algebra could have been done over the complex numbers, and the eigenvalues of the given matrix can also be defined over the complex numbers. Then from the computation of the roots above, we see that the only real eigenvalue is 1; and that there are two complex eigenvalues, namely

$$1 + \sqrt{-1} \quad \text{and} \quad 1 - \sqrt{-1}.$$

We let these eigenvalues be

$$\lambda_1 = 1, \quad \lambda_2 = 1 + \sqrt{-1}, \quad \lambda_3 = 1 - \sqrt{-1}.$$

Let

$$X = \begin{pmatrix} x \\ y \\ z \end{pmatrix}$$

be a non-zero vector. Then X is an eigenvector for A if and only if the following equations are satisfied with some eigenvalue λ:

$$x + y - z = \lambda x,$$
$$y = \lambda y,$$
$$x + z = \lambda z.$$

This system is equivalent with

$$(1 - \lambda)x + y - z = 0,$$
$$(1 - \lambda)y = 0,$$
$$x + (1 - \lambda)z = 0.$$

Case 1. $\lambda = 1$. Then the second equation will hold for any value of y. Let us put $y = 1$. From the first equation we get $z = 1$, and from the third equation we get $x = 0$. Hence we get a first eigenvector

$$X^1 = \begin{pmatrix} 0 \\ 1 \\ 1 \end{pmatrix}.$$

Case 2. $\lambda \neq 1$. Then from the second equation we must have $y = 0$. Now we solve the system arising from the first and third equations:

$$(1 - \lambda)x - z = 0,$$

$$x + (1 - \lambda)z = 0.$$

If these equations were independent, then the only solutions would be $x = z = 0$. This cannot be the case, since there must be a non-zero eigenvector with the given eigenvalue. Actually you can check directly that the second equation is equal to $(\lambda - 1)$ times the first. In any case, we give one of the variables an arbitrary value, and solve for the other. For instance, let $z = 1$. Then $x = 1/(1 - \lambda)$. Thus we get the eigenvector

$$X(\lambda) = \begin{pmatrix} 1/(1 - \lambda) \\ 0 \\ 1 \end{pmatrix}.$$

We can substitute $\lambda = \lambda_1$ and $\lambda = \lambda_2$ to get the eigenvectors with the eigenvalues λ_1 and λ_2 respectively.

In this way we have found three eigenvectors with distinct eigenvalues, namely

$$X^1, \qquad X(\lambda_1), \qquad X(\lambda_2).$$

Example 8. Find the eigenvalues and a basis for the eigenspaces of the matrix

$$\begin{pmatrix} 1 & -1 & 2 \\ -2 & 1 & 3 \\ 1 & -1 & 1 \end{pmatrix}.$$

The characteristic polynomial is

$$\begin{vmatrix} t - 1 & 1 & -2 \\ 2 & t - 1 & -3 \\ -1 & 1 & t - 1 \end{vmatrix} = (t - 1)^3 - (t - 1) - 1.$$

The eigenvalues are the roots of this cubic equation. In general it is not easy to find such roots, and this is the case in the present instance. Let $u = t - 1$. In terms of u the polynomial can be written

$$Q(u) = u^3 - u - 1.$$

From arithmetic, the only rational roots must be integers, and must divide 1, so the only possible rational roots are ± 1, which are not roots.

Hence there is no rational eigenvalue. But a cubic equation has the general shape as shown on the figure:

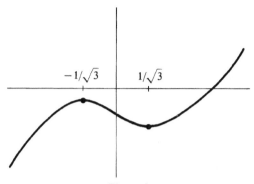

$-1/\sqrt{3}$ $1/\sqrt{3}$

Figure 1

This means that there is at least one real root. If you know calculus, then you have tools to be able to determine the relative maximum and relative minimum, you will find that the function $u^3 - u - 1$ has its relative maximum at $u = -1/\sqrt{3}$, and that $Q(-1/\sqrt{3})$ is negative. Hence there is only one real root. The other two roots are complex. This is as far as we are able to go with the means at hand. In any case, we give these roots a name, and let the eigenvalues be

$$\lambda_1, \lambda_2, \lambda_3.$$

They are all distinct.

We can, however, find the eigenvectors in terms of the eigenvalues. Let

$$X = \begin{pmatrix} x \\ y \\ z \end{pmatrix}$$

be a non-zero vector. Then X is an eigenvector if and only if $AX = \lambda X$, that is:

$$x - y + 2z = \lambda x,$$
$$-2x + y + 3z = \lambda y,$$
$$x - y + z = \lambda z.$$

This system of equations is equivalent with

$$(1 - \lambda)x - y + 2z = 0,$$
$$-2x + (1 - \lambda)y + 3z = 0,$$
$$x - y + (1 - \lambda)z = 0.$$

We give z an arbitrary value, say $z = 1$ and solve for x and y using the first two equations. Thus we must solve:

$$(\lambda - 1)x + y = 2,$$

$$2x + (\lambda - 1)y = 3.$$

Multiply the first equation by 2, the second by $(\lambda - 1)$ and subtract. Then we can solve for y to get

$$y(\lambda) = \frac{3(\lambda - 1) - 4}{(\lambda - 1)^2 - 2}.$$

From the first equation we find

$$x(\lambda) = \frac{2 - y}{\lambda - 1}.$$

Hence eigenvectors are

$$X(\lambda_1) = \begin{pmatrix} x(\lambda_1) \\ y(\lambda_1) \\ 1 \end{pmatrix}, \qquad X(\lambda_2) = \begin{pmatrix} x(\lambda_2) \\ y(\lambda_2) \\ 1 \end{pmatrix}, \qquad X(\lambda_3) = \begin{pmatrix} x(\lambda_3) \\ y(\lambda_3) \\ 1 \end{pmatrix},$$

where λ_1, λ_2, λ_3 are the three eigenvalues. This is an explicit answer to the extent that you are able to determine these eigenvalues. By machine or a computer, you can use means to get approximations to λ_1, λ_2, λ_3 which will give you corresponding approximations to the three eigenvectors. Observe that we have found here the complex eigenvectors. Let λ_1 be the real eigenvalue (we have seen that there is only one). Then from the formulas for the coordinates of $X(\lambda)$, we see that $y(\lambda)$ or $x(\lambda)$ will be real if and only if λ is real. Hence there is only one real eigenvector namely $X(\lambda_1)$. The other two eigenvectors are complex. Each eigenvector is a basis for the corresponding eigenspace.

Theorem 2.3. *Let A, B be two $n \times n$ matrices, and assume that B is invertible. Then the characteristic polynomial of A is equal to the characteristic polynomial of $B^{-1}AB$.*

Proof. By definition, and properties of the determinant,

$$\text{Det}(tI - A) = \text{Det}(B^{-1}(tI - A)B) = \text{Det}(tB^{-1}B - B^{-1}AB)$$

$$= \text{Det}(tI - B^{-1}AB).$$

This proves what we wanted.

Exercises VIII, §2

1. Let A be a diagonal matrix,

$$A = \begin{pmatrix} a_1 & 0 & \cdots & 0 \\ 0 & a_2 & \cdots & 0 \\ \vdots & \vdots & & \vdots \\ 0 & 0 & \cdots & a_n \end{pmatrix}.$$

(a) What is the characteristic polynomial of A?
(b) What are its eigenvalues?

2. Let A be a triangular matrix,

$$A = \begin{pmatrix} a_{11} & 0 & \cdots & 0 \\ a_{21} & a_{22} & \cdots & 0 \\ \vdots & \vdots & & \vdots \\ a_{n1} & a_{n2} & \cdots & a_{nn} \end{pmatrix}.$$

What is the characteristic polynomial of A, and what are its eigenvalues?

Find the characteristic polynomial, eigenvalues, and bases for the eigenspaces of the following matrices.

3. (a) $\begin{pmatrix} 1 & 2 \\ 3 & 2 \end{pmatrix}$ (b) $\begin{pmatrix} 3 & 2 \\ -1 & 0 \end{pmatrix}$

(c) $\begin{pmatrix} -2 & -7 \\ 1 & 2 \end{pmatrix}$ (d) $\begin{pmatrix} 1 & 4 \\ 2 & 3 \end{pmatrix}$

4.
(a) $\begin{pmatrix} 4 & 0 & 1 \\ -2 & 1 & 0 \\ -2 & 0 & 1 \end{pmatrix}$ (b) $\begin{pmatrix} 1 & -3 & 3 \\ 3 & -5 & 3 \\ 6 & -6 & 4 \end{pmatrix}$

(c) $\begin{pmatrix} 3 & 1 & 1 \\ 2 & 4 & 2 \\ 1 & 1 & 3 \end{pmatrix}$ (d) $\begin{pmatrix} 1 & 2 & 2 \\ 1 & 2 & -1 \\ -1 & 1 & 4 \end{pmatrix}$

5. Find the eigenvalues and eigenvectors of the following matrices. Show that the eigenvectors form a 1-dimensional space.

(a) $\begin{pmatrix} 2 & -1 \\ 1 & 0 \end{pmatrix}$ (b) $\begin{pmatrix} 1 & 1 \\ 0 & 1 \end{pmatrix}$ (c) $\begin{pmatrix} 2 & 0 \\ 1 & 2 \end{pmatrix}$ (d) $\begin{pmatrix} 2 & -3 \\ 1 & -1 \end{pmatrix}$

6. Find the eigenvalues and eigenvectors of the following matrices. Show that the eigenvectors form a 1-dimensional space.

(a) $\begin{pmatrix} 1 & 1 & 1 \\ 0 & 1 & 1 \\ 0 & 0 & 1 \end{pmatrix}$ (b) $\begin{pmatrix} 1 & 1 & 0 \\ 0 & 1 & 1 \\ 0 & 0 & 1 \end{pmatrix}$

7. Find the eigenvalues and a basis for the eigenspaces of the following matrices.

(a) $\begin{pmatrix} 0 & 1 & 0 & 0 \\ 0 & 0 & 1 & 0 \\ 0 & 0 & 0 & 1 \\ 1 & 0 & 0 & 0 \end{pmatrix}$ (b) $\begin{pmatrix} -1 & 0 & 1 \\ -1 & 3 & 0 \\ -4 & 13 & -1 \end{pmatrix}$

8. Find the eigenvalues and a basis for the eigenspaces for the following matrices.

(a) $\begin{pmatrix} 2 & 4 \\ 5 & 3 \end{pmatrix}$ (b) $\begin{pmatrix} 1 & 2 \\ 2 & -2 \end{pmatrix}$ (c) $\begin{pmatrix} 3 & 2 \\ -2 & 3 \end{pmatrix}$

(d) $\begin{pmatrix} -1 & 2 & 2 \\ 2 & 2 & 2 \\ -3 & -6 & -6 \end{pmatrix}$ (e) $\begin{pmatrix} 3 & 2 & 1 \\ 0 & 1 & 2 \\ 0 & 1 & -1 \end{pmatrix}$ (f) $\begin{pmatrix} -1 & 4 & -2 \\ -3 & 4 & 0 \\ -3 & 1 & 3 \end{pmatrix}$

9. Let V be an n-dimensional vector space and assume that the characteristic polynomial of a linear map $A: V \to V$ has n distinct roots. Show that V has a basis consisting of eigenvectors of A.

10. Let A be a square matrix. Shows that the eigenvalues of $^t A$ are the same as those of A.

11. Let A be an invertible matrix. If λ is an eigenvalue of A show that $\lambda \neq 0$ and that λ^{-1} is an eigenvalue of A^{-1}.

12. Let V be the space generated over $\mathbf{R}$ by the two functions $\sin t$ and $\cos t$. Does the derivative (viewed as a linear map of V into itself) have any non-zero eigenvectors in V? If so, which?

13. Let D dénote the derivative which we view as a linear map on the space of differentiable functions. Let k be an integer $\neq 0$. Show that the functions $\sin kx$ and $\cos kx$ are eigenvectors for D^2. What are the eigenvalues?

14. Let $A: V \to V$ be a linear map of V into itself, and let $\{v_1, \dots, v_n\}$ be a basis of V consisting of eigenvectors having distinct eigenvalues $c_1, \dots, c_n$. Show that any eigenvector v of A in V is a scalar multiple of some v_i.

15. Let A, B be square matrices of the same size. Show that the eigenvalues of AB are the same as the eigenvalues of BA.

VIII, §3. Eigenvalues and Eigenvectors of Symmetric Matrices

We shall give two proofs of the following theorem.

Theorem 3.1. *Let A be a symmetric $n \times n$ real matrix. Then there exists a non-zero real eigenvector for A.*

One of the proofs will use the complex numbers, and the other proof will use calculus. Let us start with the calculus proof.

Define the function

$$f(X) = {}^tXAX \qquad \text{for} \qquad X \in \mathbf{R}^n.$$

Such a function f is called the **quadratic form associated with** A. If ${}^tX = (x_1, \ldots, x_n)$ is written in terms of coordinates, and $A = (a_{ij})$ then

$$f(X) = \sum_{i, j=1}^{n} a_{ij} x_i x_j.$$

Example. Let

$$A = \begin{pmatrix} 3 & -1 \\ -1 & 2 \end{pmatrix}.$$

Let ${}^tX = (x, y)$. Then

$${}^tXAX = (x, y) \begin{pmatrix} 3 & -1 \\ -1 & 2 \end{pmatrix} \begin{pmatrix} x \\ y \end{pmatrix} = 3x^2 - 2xy + 2y^2.$$

More generally, let

$$A = \begin{pmatrix} a & b \\ b & d \end{pmatrix}.$$

Then

$$(x, y) \begin{pmatrix} a & b \\ b & d \end{pmatrix} \begin{pmatrix} x \\ y \end{pmatrix} = ax^2 + 2bxy + dy^2.$$

Example. Suppose we are given a quadratic expression

$$f(x, y) = 3x^2 + 5xy - 4y^2.$$

Then it is the quadratic form associated with the symmetric matrix

$$A = \begin{pmatrix} 3 & \frac{5}{2} \\ \frac{5}{2} & -4 \end{pmatrix}.$$

In many applications, one wants to find a maximum for such a function f on the unit sphere. Recall that the **unit sphere** is the set of all points X such that $\|X\| = 1$, where $\|X\| = \sqrt{X \cdot X}$. It is shown in analysis courses that a continuous function f as above necessarily has a maximum on the sphere. A **maximum** on the unit sphere is a point P such that $\|P\| = 1$ and

$$f(P) \geq f(X) \qquad \text{for all } X \text{ with } \|X\| = 1.$$

The next theorem relates this problem with the problem of finding eigenvectors.

Theorem 3.2. *Let A be a real symmetric matrix, and let $f(X) = {}^t X A X$ be the associated quadratic form. Let P be a point on the unit sphere such that $f(P)$ is a maximum for f on the sphere. Then P is an eigenvector for A. In other words, there exists a number λ such that $AP = \lambda P$.*

Proof. Let W be the subspace of $\mathbf{R}^n$ orthogonal to P, that is $W = P^\perp$. Then $\dim W = n - 1$. For any element $w \in W$, $\|w\| = 1$, define the curve

$$C(t) = (\cos t)P + (\sin t)w.$$

The directions of unit vectors $w \in W$ are the directions tangent to the sphere at the point P, as shown on the figure.

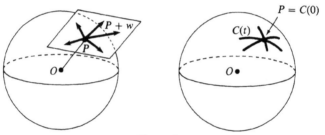

Figure 2

The curve $C(t)$ lies on the sphere because $\|C(t)\| = 1$, as you can verify at once by taking the dot product $C(t) \cdot C(t)$, and using the hypothesis that $P \cdot w = 0$. Furthermore, $C(0) = P$, so $C(t)$ is a curve on the sphere passing through P. We also have the derivative

$$C'(t) = (-\sin t)P + (\cos t)w,$$

and so $C'(0) = w$. Thus the direction of the curve is in the direction of w, and is perpendicular to the sphere at P because $w \cdot P = 0$. Consider the function

$$g(t) = f(C(t)) = C(t) \cdot AC(t).$$

Using coordinates, and the rule for the derivative of a product which applies in this case (as you might know from calculus), you find the derivative:

$$g'(t) = C'(t) \cdot AC(t) + C(t) \cdot AC'(t)$$
$$= 2C'(t) \cdot AC(t),$$

because A is symmetric. Since $f(P)$ is a maximum and $g(0) = f(P)$, it follows that $g'(0) = 0$. Then we obtain:

$$O = g'(0) = 2C'(0) \cdot AC(0) = 2w \cdot AP.$$

Hence AP is perpendicular to W for all $w \in W$. But $W^\perp$ is the 1-dimensional space generated by P. Hence there is a number λ such that $AP = \lambda P$, thus proving the theorem.

Corollary 3.3. *The maximum value of f on the unit sphere is equal to the largest eigenvalue of A.*

Proof. Let λ be any eigenvalue and let P be an eigenvector on the unit sphere, so $\|P\| = 1$. Then

$$f(P) = {}^tPAP = {}^tP\lambda P = \lambda\, {}^tPP = \lambda.$$

Thus the value of f at an eigenvector on the unit sphere is equal to the eigenvalue. Theorem 3.2 tells us that the maximum of f on the unit sphere occurs at an eigenvector. Hence the maximum of f on the unit sphere is equal to the largest eigenvalue, as asserted.

Example. Let $f(x, y) = 2x^2 - 3xy + y^2$. Let A be the symmetric matrix associated with f. Find the eigenvectors of A on the unit circle, and find the maximum of f on the unit circle.

First we note that f is the quadratic form associated with the matrix

$$A = \begin{pmatrix} 2 & -\frac{3}{2} \\ -\frac{3}{2} & 1 \end{pmatrix}.$$

By Theorem 3.2 a maximum must occur at an eigenvector, so we first find the eigenvalues and eigenvectors.

The characteristic polynomial is the determinant

$$\begin{vmatrix} t - 2 & \frac{3}{2} \\ \frac{3}{2} & t - 1 \end{vmatrix} = t^2 - 3t - \tfrac{1}{4}.$$

Then the eigenvalues are

$$\lambda = \frac{3 \pm \sqrt{10}}{2}.$$

For the eigenvectors, we must solve

$$2x - \tfrac{3}{2}y = \lambda x,$$
$$-\tfrac{3}{2}x + y = \lambda y.$$

Putting $x = 1$ this gives the possible eigenvectors

$$X(\lambda) = \begin{pmatrix} 1 \\ \frac{2}{3}(2 - \lambda) \end{pmatrix}.$$

Thus there are two such eigenvectors, up to non-zero scalar multiples. The eigenvectors lying on the unit circle are therefore

$$P(\lambda) = \frac{X(\lambda)}{\|X(\lambda)\|} \quad \text{with} \quad \lambda = \frac{3 + \sqrt{10}}{2} \quad \text{and} \quad \lambda = \frac{3 - \sqrt{10}}{2}.$$

By Corollary 3.3 the maximum is the point with the bigger eigenvalue, and must therefore be the point

$$P(\lambda) \quad \text{with} \quad \lambda = \frac{3 + \sqrt{10}}{2}.$$

The maximum value of f on the unit circle is $(3 + \sqrt{10})/2$.

By the same token, the minimum value of f on the unit circle is $(3 - \sqrt{10})/2$.

We shall now use the complex numbers $\mathbf{C}$ for the second proof. A fundamental property of complex numbers is that every non-constant polynomial with complex coefficients has a root (a zero) in the complex numbers. Therefore the characteristic polynomial of A has a complex root λ, which is a priori a complex eigenvalue, with a complex eigenvector.

Theorem 3.4. *Let A be a real symmetric matrix and let λ be an eigenvalue in $\mathbf{C}$. Then λ is real. If $Z \neq O$ is a complex eigenvector with eigenvalue λ, and $Z = X + iY$ where $X, Y \in \mathbf{R}^n$, then both X, Y are real eigenvectors of A with eigenvalue λ, and X or $Y \neq O$.*

Proof. Let $Z = {}^t(z_1, \ldots, z_n)$ with complex coordinates z_i. Then

$$Z \cdot \bar{Z} = \bar{Z} \cdot Z = {}^t\bar{Z}Z = \bar{z}_1 z_1 + \cdots + \bar{z}_n z_n = |z_1|^2 + \cdots + |z_n|^2 > 0.$$

By hypothesis, we have $AZ = \lambda Z$. Then

$${}^t\bar{Z}AZ = {}^t\bar{Z}\lambda Z = \lambda {}^t\bar{Z}Z.$$

The transpose of a 1×1 matrix is equal to itself, so we also get

$${}^tZ'A\bar{Z} = {}^tZA\bar{Z} = \lambda {}^t\bar{Z}Z.$$

But $\overline{AZ} = \overline{A}\overline{Z} = A\overline{Z}$ and $\overline{AZ} = \overline{\lambda Z} = \overline{\lambda}\overline{Z}$. Therefore

$$\lambda {}^t\overline{Z}Z = \overline{\lambda}\, {}^tZ\overline{Z}.$$

Since ${}^tZ\overline{Z} \neq 0$ it follows that $\lambda = \overline{\lambda}$, so λ is real.

Now from $AZ = \lambda Z$ we get

$$AX + iAY = \lambda X + i\lambda Y,$$

and since A, X, Y, are real it follows that $AX = \lambda X$ and $AY = \lambda Y$. This proves the theorem.

Exercises VIII, §3

1. Find the eigenvalues of the following matrices, and the maximum value of the associated quadratic forms on the unit circle.

(a) $\begin{pmatrix} 2 & -1 \\ -1 & 2 \end{pmatrix}$ (b) $\begin{pmatrix} 1 & 1 \\ 1 & 0 \end{pmatrix}$

2. Same question, except find the maximum on the unit sphere.

(a) $\begin{pmatrix} 1 & -1 & 0 \\ -1 & 2 & -1 \\ 0 & -1 & 1 \end{pmatrix}$ (b) $\begin{pmatrix} 2 & -1 & 0 \\ -1 & 2 & -1 \\ 0 & -1 & 2 \end{pmatrix}$

3. Find the maximum and minimum of the function

$$f(x, y) = 3x^2 + 5xy - 4y^2$$

on the unit circle.

VIII, §4. Diagonalization of a Symmetric Linear Map

In this section we give an application of the existence of eigenvectors as proved in §3. Since we shall do an induction, instead of working with $\mathbf{R}^n$ we have to start with a formulation dealing with a vector space in which coordinates have not yet been chosen.

So let V be a vector space of dimension n over $\mathbf{R}$, with a positive definite scalar product $\langle v, w \rangle$ for v, $w \in V$. Let

$$A: V \to V$$

be a linear map. We shall say that A is **symmetric (with respect to the scalar product)** if we have the relation

$$\langle Av, w \rangle = \langle v, Aw \rangle$$

for all $v, w \in V$.

Example. Suppose $V = \mathbf{R}^n$ and that the scalar product is the usual dot product between vectors. A linear map A is represented by a matrix, and we use the same letter A to denote this matrix. Then for all v, $w \in \mathbf{R}^n$ we have

$$\langle Av, w \rangle = {}^t w A v$$

if we view v, w as column vectors. Since ${}^t w A v$ is a 1×1 matrix, it is equal to its transpose. Thus we have the formula

$${}^t w A v = {}^t v {}^t A w,$$

or in terms of the $\langle \ , \ \rangle$ notation,

$$\langle Av, w \rangle = \langle v, {}^t A w \rangle.$$

The condition that A is symmetric as a linear map with respect to the scalar product is by definition $\langle Av, w \rangle = \langle v, Aw \rangle$, or in terms of matrix multiplication

$${}^t w A v = {}^t v A w.$$

Comparing with the previous formula, this means that $A = {}^t A$. Thus we find:

> Let A be an $n \times n$ matrix, and L_A the associated linear map on $\mathbf{R}^n$. Then L_A is symmetric with respect to the scalar product if and only if A is a symmetric matrix.

If V is a general vector space of dimension n with a positive definite scalar product, and $A: V \to V$ is a linear map of V into itself, then there is a unique linear

$${}^t A: V \to V$$

satisfying the formula

$$\boxed{\langle Av, w \rangle = \langle v, {}^t A w \rangle.}$$

We have just seen this when $V = \mathbf{R}^n$. In general, we pick an orthonormal basis of V which allows us to identify V with $\mathbf{R}^n$, and to identify the scalar product with the ordinary dot product. Then ${}^t A$ (as a linear map) coincides with the transpose of the matrix representing A.

We can now say that a linear map $A: V \to V$ is symmetric if and only if it is equal to its own transpose. When V is identified with $\mathbf{R}^n$ by using an orthonormal basis, this means that the matrix representing A is equal to its transpose, in other words, the matrix is symmetric.

We can reformulate Theorem 3.1 as follows:

Let V be a finite dimensional vector space with a positive definite scalar product. Let $A: V \to V$ be a symmetric linear map. Then A has a non-zero eigenvector.

Let W be a subspace of V, and let $A: V \to V$ be a symmetric linear map. We say that W is **stable under** A if $A(W) \subset W$, that is for all $u \in W$ we have $Au \in W$.

We note that if W is stable under A then its orthogonal complement $W^\perp$ is also stable under A.

Proof. Let $w \in W^\perp$. Then for all $u \in W$ we have

$$\langle Aw, u \rangle = \langle w, Au \rangle = 0$$

because $Au \in W$ and $w \in W^\perp$. Hence $Aw \in W^\perp$, thus proving the assertion.

Theorem 4.1. *Let V be a finite dimensional vector space over the real numbers, of dimension $n > 0$, and with a positive definite scalar product. Let*

$$A: V \to V$$

be a linear map, symmetric with respect to the scalar product. Then V has an orthonormal basis consisting of eigenvectors.

Proof. By Theorem 3.1, there exists a non-zero eigenvector P for A. Let W be the one-dimensional space generated by P. Then W is stable under A. By the above remark, $W^\perp$ is also stable under A and is a vector space of dimension $n - 1$. We may then view A as giving a symmetric linear map of $W^\perp$ into itself. We can then repeat the procedure We put $P = P_1$, and by induction we can find a basis $\{P_2, \ldots, P_n\}$ of $W^\perp$ consisting of eigenvectors. Then

$$\{P_1, P_2, \ldots, P_n\}$$

is an orthogonal basis of V consisting of eigenvectors. We divide each vector by its norm to get an orthonormal basis, as desired.

If $\{e_1, \ldots, e_n\}$ is an orthonormal basis of V such that each e_i is an eigenvector, then the matrix of A with respect to this basis is diagonal, and the diagonal elements are precisely the eigenvalues:

$$\begin{pmatrix} \lambda_1 & 0 & \cdots & 0 \\ 0 & \lambda_2 & \cdots & 0 \\ \vdots & \vdots & & \vdots \\ 0 & 0 & \cdots & \lambda_n \end{pmatrix}.$$

In such a simple representation, the effect of A then becomes much clearer than when A is represented by a more complicated matrix with respect to another basis.

Example. We give an application to linear differential equations. Let A be an $n \times n$ symmetric real matrix. We want to find the solutions in $\mathbf{R}^n$ of the differential equation

$$\frac{dX(t)}{dt} = AX(t),$$

where

$$X(t) = \begin{pmatrix} x_1(t) \\ \vdots \\ x_n(t) \end{pmatrix}$$

is given in terms of coordinates which are functions of t, and

$$\frac{dX(t)}{dt} = \begin{pmatrix} dx_1/dt \\ \vdots \\ dx_n/dt \end{pmatrix}.$$

Writing this equation in terms of arbitrary coordinates is messy. So let us forget at first about coordinates, and view $\mathbf{R}^n$ as an n-dimensional vector space with a positive definite scalar product. We choose an orthonormal basis of V (usually different from the original basis) consisting of eigenvectors of A. Now *with respect to this new basis*, we can identify V with $\mathbf{R}^n$ with new coordinates which we denote by $y_1, \ldots, y_n$.

With respect to these new coordinates, the matrix of the linear map L_A is

$$\begin{pmatrix} \lambda_1 & 0 & \cdots & 0 \\ 0 & \lambda_2 & \cdots & 0 \\ \vdots & \vdots & \ddots & \vdots \\ 0 & 0 & \cdots & \lambda_n \end{pmatrix}$$

where $\lambda_1, \ldots, \lambda_n$ are the eigenvalues. But in terms of these more convenient coordinates, our differential equation simply reads

$$\frac{dy_1}{dt} = \lambda_1 y_1, \quad \ldots, \quad \frac{dy_n}{dt} = \lambda_n y_n.$$

Thus the most general solution is of the form

$$y_i(t) = c_i e^{\lambda_i t} \quad \text{with some constant } c_i.$$

The moral of this example is that one should not select a basis too quickly, and one should use as often as possible a notation without coordinates, until a choice of coordinates becomes imperative to make the solution of a problem simpler.

Exercises VIII, §4

1. Suppose that A is a diagonal $n \times n$ matrix. For any $X \in \mathbf{R}^n$, what is ${}^t X A X$ in terms of the coordinates of X and the diagonal elements of A?

2. Let

$$A = \begin{pmatrix} \lambda_1 & 0 & \cdots & 0 \\ 0 & \lambda_2 & \cdots & 0 \\ \vdots & & & \vdots \\ 0 & 0 & \cdots & \lambda_n \end{pmatrix}$$

be a diagonal matrix with $\lambda_1 \geq 0, \ldots, \lambda_n \geq 0$. Show that there exists an $n \times n$ diagonal matrix B such that $B^2 = A$.

3. Let V be a finite dimensional vector space with a positive definite scalar product. Let $A: V \to V$ be a symmetric linear map. We say that A is **positive definite** if $\langle Av, v \rangle > 0$ for all $v \in V$ and $v \neq 0$. Prove:
   (a) If A is positive definite, then all eigenvalues are > 0.
   (b) If A is positive definite, then there exists a symmetric linear map B such that $B^2 = A$ and $BA = AB$. What are the eigenvalues of B? [*Hint:* Use a basis of V consisting of eigenvectors.]

4. We say that A is **positive semidefinite** if $\langle Av, v \rangle \geq 0$ for all $v \in V$. Prove the analogues of (a), (b) of Exercise 3 when A is only assumed semidefinite. Thus the eigenvalues are ≥ 0, and there exists a symmetric linear map B such that $B^2 = A$.

5. Assume that A is symmetric positive definite. Show that A^2 and A^{-1} are symmetric positive definite.

6. Let $U: \mathbf{R}^n \to \mathbf{R}^n$ be a linear map, and let $\langle \, , \, \rangle$ denote the usual scalar (dot) product. Show that the following conditions are equivalent:
   (i) $\|Uv\| = \|v\|$ for all $v \in \mathbf{R}^n$.
   (ii) $\langle Uv, Uw \rangle = \langle v, w \rangle$ for all $v, w \in \mathbf{R}^n$.
   (iii) U is invertible, and ${}^tU = U^{-1}$.
   [*Hint:* For (ii), use the identity

$$\langle (v + w), (v + w) \rangle - \langle (v - w), (v - w) \rangle = 4 \langle v, w \rangle,$$

and similarly with a U in front of each vector.] When U satisfies any one (and hence all) of these conditions, then U is called **unitary**. The first condition says that U preserves the norm, and the second says that U preserves the scalar product.

7. Let $A: \mathbf{R}^n \to \mathbf{R}^n$ be an invertible linear map.
   (i) Show that ${}^tAA$ is symmetric positive definite.
   (ii) By Exercise 3b, there is a symmetric positive definite B such that $B^2 = {}^tAA$. Let $U = AB^{-1}$. Show that U is unitary.
   (iii) Show that $A = UB$.

8. Let B be symmetric positive definite and also unitary. Show that $B = I$.

Appendix. Complex Numbers

The **complex numbers C** are a set of objects which can be added and multiplied, the sum and product of two complex numbers being also complex numbers, and satisfying the following conditions:

(1) Every real number is a complex number, and if α, β are real numbers, then their sum and product as complex numbers are the same as their sum and product as real numbers.

(2) There is a complex number denoted by i such that $i^2 = -1$.

(3) Every complex number can be written uniquely in the form $a + bi$, where a, b are real numbers.

(4) The ordinary laws of arithmetic concerning addition and multiplication are satisfied. We list these laws:

If α, β, γ are complex numbers, then

$$(\alpha + \beta) + \gamma = \alpha + (\beta + \gamma) \qquad \text{and} \qquad (\alpha\beta)\gamma = \alpha(\beta\gamma).$$

We have $\alpha(\beta + \gamma) = \alpha\beta + \alpha\gamma$ and $(\beta + \gamma)\alpha = \beta\alpha + \gamma\alpha$.
We have $\alpha\beta = \beta\alpha$ and $\alpha + \beta = \beta + \alpha$.
If 1 is the real number one, then $1\alpha = \alpha$.
If 0 is the real number zero, then $0\alpha = 0$.
We have $\alpha + (-1)\alpha = 0$.

We shall now draw consequences of these properties. If we write

$$\alpha = a_1 + a_2 i \qquad \text{and} \qquad \beta = b_1 + b_2 i,$$

then

$$\alpha + \beta = a_1 + a_2 i + b_1 + b_2 i = a_1 + b_1 + (a_2 + b_2)i.$$

If we call a_1 the **real part**, or real component of α, and a_2 its **imaginary part**, or imaginary component, then we see that addition is carried out componentwise. The real part and imaginary part of α are denoted by $\text{Re}(\alpha)$ and $\text{Im}(\alpha)$ respectively.

We have

$$\alpha\beta = (a_1 + a_2 i)(b_1 + b_2 i) = a_1 b_1 - a_2 b_2 + (a_2 b_1 + a_1 b_2)i.$$

Let $\alpha = a + bi$ be a complex number with a, b real. We define

$$\bar{\alpha} = a - bi$$

and call $\bar{\alpha}$ the complex conjugate, or simply **conjugate**, of α. Then

$$\alpha\bar{\alpha} = a^2 + b^2.$$

If $\alpha = a + bi$ is $\neq 0$, and if we let

$$\lambda = \frac{\bar{\alpha}}{a^2 + b^2},$$

then $\alpha\lambda = \lambda\alpha = 1$, as we see immediately. The number λ above is called the **inverse** of α and is denoted by α^{-1}, or $1/\alpha$. We note that it is the only complex number z such that $z\alpha = 1$, because if this equation is satisfied, we multiply it by λ on the right to find $z = \lambda$. If α, β are complex numbers, we often write β/α instead of $\alpha^{-1}\beta$ or $\beta\alpha^{-1}$. We see that we can divide by complex numbers $\neq 0$.

We have the rules

$$\overline{\alpha\beta} = \bar{\alpha}\bar{\beta}, \qquad \overline{\alpha + \beta} = \bar{\alpha} + \bar{\beta}, \qquad \bar{\bar{\alpha}} = \alpha.$$

These follow at once from the definitions of addition and multiplication.

We define the **absolute value** of a complex number $\alpha = a + bi$ to be

$$|\alpha| = \sqrt{a^2 + b^2}.$$

If we think of α as a point in the plane (a, b), then $|\alpha|$ is the length of the line segment from the origin to α. In terms of the absolute value, we can write

$$\alpha^{-1} = \frac{\bar{\alpha}}{|\alpha|^2}$$

provided $\alpha \neq 0$. Indeed, we observe that $|\alpha|^2 = \alpha\bar{\alpha}$. Note also that $|\alpha| = |\bar{\alpha}|$.

The absolute value satisfies properties analogous to those satisfied by the absolute value of real numbers:

$$|\alpha| \geq 0 \text{ and } = 0 \text{ if and only if } \alpha = 0.$$

$$|\alpha\beta| = |\alpha|\,|\beta|$$

$$|\alpha + \beta| \leq |\alpha| + |\beta|.$$

The first assertion is obvious. As to the second, we have

$$|\alpha\beta|^2 = \alpha\beta\bar{\alpha}\bar{\beta} = |\alpha|^2|\beta|^2.$$

Taking the square root, we conclude that $|\alpha|\,|\beta| = |\alpha\beta|$. Next, we have

$$|\alpha + \beta|^2 = (\alpha + \beta)\overline{(\alpha + \beta)} = (\alpha + \beta)(\bar{\alpha} + \bar{\beta})$$

because $\alpha\bar{\beta} = \overline{\bar{\beta}\alpha}$. However, we have

$$2\,\mathrm{Re}(\beta\bar{\alpha}) \leq 2|\beta\bar{\alpha}|$$

because the real part of a complex number is $\leq$ its absolute value. Hence

$$|\alpha + \beta|^2 \leq |\alpha|^2 + 2|\beta\bar{\alpha}| + |\beta|^2$$

$$\leq |\alpha|^2 + 2|\beta|\,|\alpha| + |\beta|^2$$

$$= (|\alpha| + |\beta|)^2.$$

Taking the square root yields the final property.

Let $z = x + iy$ be a complex number $\neq 0$. Then $z/|z|$ has absolute value 1.

The main advantage of working with complex numbers rather than real numbers is that every non-constant polynomial with complex coefficients has a root in **C**. This is proved in more advanced courses in analysis. For instance, a quadratic equation

$$ax^2 + bx + c = 0,$$

with $a \neq 0$ has the roots

$$x = \frac{-b \pm \sqrt{b^2 - 4ac}}{2a}.$$

If $b^2 - 4ac$ is positive, then the roots are real. If $b^2 - 4ac$ is negative, then the roots are complex. The proof for the quadratic formula uses only the basic arithmetic of addition, multiplication, and division. Namely, we complete the square to see that

$$ax^2 + bx + c = a\left(x + \frac{b}{2a}\right)^2 - \frac{b^2}{4a} + c = a\left(x + \frac{b}{2a}\right)^2 - \frac{b^2 - 4ac}{4a}.$$

Then we solve

$$\left(x + \frac{b}{2a}\right)^2 = \frac{b^2 - 4ac}{4a^2},$$

take the square root, and finally get the desired value for x.

Application to vector spaces

To define the notion of a vector space, we need first the notion of scalars. And the only facts we need about scalars are those connected with addition, multiplication, and division by non-zero elements. These basic operations of arithmetic are all satisfied by the complex numbers. Therefore we can do the basic theory of vector spaces over the complex numbers. We have the same theorems about linear combinations, matrices, row rank, column rank, dimension, determinants, characteristic polynomials, eigenvalues.

The only basic difference (and it is slight) comes when we deal with the dot product. If $Z = (z_1, \ldots, z_n)$ and $W = (w_1, \ldots, w_n)$ are n-tuples in **C**^{n}, then their dot product is as before

$$Z \cdot W = z_1 w_1 + \cdots + z_n w_n.$$

But observe that even if $Z \neq O$, then $Z \cdot Z$ may be 0. For instance, let $Z = (1, i)$ in **C**^{2}. Then

$$Z \cdot Z = 1 + i^2 = 1 - 1 = 0.$$

Hence the dot product is not positive definite.

To remedy this, one defines a product which is called **hermitian** and is almost equal to the dot product, but contains a complex conjugate. That is we define $\bar{W} = (\bar{w}_1, \ldots, \bar{w}_n)$ and

$$\langle Z, W \rangle = Z \cdot \bar{W} = z_1 \bar{w}_1 + \cdots + z_n \bar{w}_n,$$

so we put a *complex conjugate on the coordinates of W.* Then

$$\langle Z, Z \rangle = Z \cdot \bar{Z} = z_1 \bar{z}_1 + \cdots + z_n \bar{z}_n = |z_1|^2 + \cdots + |z_n|^2.$$

Hence once again, if $Z \neq O$, then some coordinate $z_i \neq 0$, so the sum on the right is $\neq 0$ and $\langle Z, Z \rangle > 0$.

If α is a complex number, then from the definition we see that

$$\langle \alpha Z, W \rangle = \alpha \langle Z, W \rangle \qquad \text{but} \qquad \langle Z, \alpha W \rangle = \bar{\alpha} \langle Z, W \rangle.$$

Thus a complex conjugate appears in the second formula. We still have the formulas expressing additivity

$$\langle Z_1 + Z_2, W \rangle = \langle Z_1, W \rangle + \langle Z_2, W \rangle$$

and

$$\langle Z, W_1 + W_2 \rangle = \langle Z, W_1 \rangle + \langle Z, W_2 \rangle.$$

We then say that the hermitian product is **linear** in the first variable and **antilinear** in the second variable. Note that instead of commutativity of the hermitian product, we have the formula

$$\langle Z, W \rangle = \overline{\langle W, Z \rangle}.$$

If Z, W are real vectors, then the hermitian product $\langle Z, W \rangle$ is the same as the dot product.

One can then develop the Gram–Schmidt orthogonalization process just as before using the hermitian product rather than the dot product.

In the application of this chapter we did not need the hermitian product. All we needed was that a complex $n \times n$ matrix A has an eigenvalue, and that the eigenvalues are the roots of the characteristic polynomial

$$\det(tI - A).$$

As mentioned before, a non-constant polynomial with complex coefficients always has a root in the complex numbers, so A always has an eigenvalue in **C**. In the text, we showed that when A is real symmetric, then such eigenvalues must in fact be real.

Answers to Exercises

I, §1, p. 8

	$A + B$	$A - B$	$3A$	$-2B$
1.	$(1, 0)$	$(3, -2)$	$(6, -3)$	$(2, -2)$
2.	$(-1, 7)$	$(-1, -1)$	$(-3, 9)$	$(0, -8)$
3.	$(1, 0, 6)$	$(3, -2, 4)$	$(6, -3, 15)$	$(2, -2, -2)$
4.	$(-2, 1, -1)$	$(0, -5, 7)$	$(-3, -6, 9)$	$(2, -6, 8)$
5.	$(3\pi, 0, 6)$	$(-\pi, 6, -8)$	$(3\pi, 9, -3)$	$(-4\pi, 6, -14)$
6.	$(15 + \pi, 1, 3)$	$(15 - \pi, -5, 5)$	$(45, -6, 12)$	$(-2\pi, -6, 2)$

I, §2, p. 12

1. No   **2.** Yes   **3.** No   **4.** Yes   **5.** No   **6.** Yes   **7.** Yes   **8.** No

I, §3, p. 15

1. (a) 5   (b) 10   (c) 30   (d) 14   (e) $\pi^2 + 10$ (f) 245

2. (a) -3 (b) 12   (c) 2   (d) -17 (e) $2\pi^2 - 16$ (f) $15\pi - 10$

4. (b) and (d)

I, §4, p. 29

1. (a) $\sqrt{5}$ (b) $\sqrt{10}$ (c) $\sqrt{30}$ (d) $\sqrt{14}$ (e) $\sqrt{10 + \pi^2}$ (f) $\sqrt{245}$

2. (a) $\sqrt{2}$ (b) 4 (c) $\sqrt{3}$ (d) $\sqrt{26}$ (e) $\sqrt{58 + 4\pi^2}$ (f) $\sqrt{10 + \pi^2}$

3. (a) $(\frac{3}{2}, -\frac{3}{2})$ (b) (0, 3) (c) $(-\frac{2}{3}, \frac{2}{3}, \frac{2}{3})$ (d) $(\frac{17}{26}, -\frac{51}{26}, \frac{34}{13})$

(e) $\dfrac{\pi^2 - 8}{2\pi^2 + 29}$ $(2\pi, -3, 7)$ (f) $\dfrac{15\pi - 10}{10 + \pi^2}$ $(\pi, 3, -1)$

4. (a) $(-\frac{6}{5}, \frac{3}{5})$ (b) $(-\frac{6}{5}, \frac{18}{5})$ (c) $(\frac{2}{15}, -\frac{1}{15}, \frac{1}{3})$ (d) $-\frac{17}{14}(-1, -2, 3)$

(e) $\dfrac{2\pi^2 - 16}{\pi^2 + 10}$ $(\pi, 3, -1)$ (f) $\dfrac{3\pi - 2}{49}$ $(15, -2, 4)$

5. (a) $\dfrac{-1}{\sqrt{5}\sqrt{34}}$ (b) $\dfrac{-2}{\sqrt{5}}$ (c) $\dfrac{10}{\sqrt{14}\sqrt{35}}$ (d) $\dfrac{13}{\sqrt{21}\sqrt{11}}$ (e) $\dfrac{-1}{\sqrt{12}}$

6. (a) $\dfrac{35}{\sqrt{41 \cdot 35}}, \dfrac{6}{\sqrt{41 \cdot 6}}, 0$ (b) $\dfrac{1}{\sqrt{17 \cdot 26}}, \dfrac{16}{\sqrt{41 \cdot 17}}, \dfrac{25}{\sqrt{26 \cdot 41}}$

7. Let us dot the sum

$$c_1 A_1 + \cdots + c_r A_r = O$$

with A_i. We find

$$c_1 A_1 \cdot A_i + \cdots + c_i A_i \cdot A_i + \cdots + c_r A_r \cdot A_i = O \cdot A_i = 0.$$

Since $A_j \cdot A_i = 0$ if $j \neq i$ we find

$$c_i A_i \cdot A_i = 0.$$

But $A_i \cdot A_i \neq 0$ by assumption. Hence $c_i = 0$, as was to be shown.

8. (a) $\|A + B\|^2 + \|A - B\|^2 = (A + B) \cdot (A + B) + (A - B) \cdot (A - B)$

$$= A^2 + 2A \cdot B + B^2 + A^2 - 2A \cdot B + B^2$$

$$= 2A^2 + 2B^2 = 2\|A\|^2 + 2\|B\|^2$$

9. $\|A - B\|^2 = A^2 - 2A \cdot B + B^2 = \|A\|^2 - 2\|A\| \|B\| \cos \theta + \|B\|^2$

I, §5, p. 34

1. (a) Let $A = P_2 - P_1 = (-5, -2, 3)$. Parametric representation of the line is $X(t) = P_1 + tA = (1, 3, -1) + t(-5, -2, 3)$.
(b) $(-1, 5, 3) + t(-1, -1, 4)$

2. $X = (1, 1, -1) + t(3, 0, -4)$ **3.** $X = (-1, 5, 2) + t(-4, 9, 1)$

4. (a) $(-\frac{3}{2}, 4, \frac{1}{2})$ (b) $(-\frac{2}{3}, \frac{11}{3}, 0)$, $(-\frac{7}{3}, \frac{13}{3}, 1)$ (c) $(0, \frac{17}{5}, -\frac{2}{5})$ (d) $(-1, \frac{19}{5}, \frac{1}{5})$

5. $P + \frac{1}{2}(Q - P) = \dfrac{P + Q}{2}$

I, §6, p. 40

1. The normal vectors $(2, 3)$ and $(5, -5)$ are not perpendicular because their dot product $10 - 15 = -5$ is not 0.

2. The normal vectors are $(-m, 1)$ and $(-m', 1)$, and their dot product is $mm' + 1$. The vectors are perpendicular if and only if this dot product is 0, which is equivalent with $mm' = -1$.

3. $y = x + 8$ 4. $4y = 5x - 7$ 6. (c) and (d)

7. (a) $x - y + 3z = -1$ (b) $3x + 2y - 4z = 2\pi + 26$ (c) $x - 5z = -33$

8. (a) $2x + y + 2z = 7$ (b) $7x - 8y - 9z = -29$ (c) $y + z = 1$

9. $(3, -9, -5)$, $(1, 5, -7)$ (Others would be constant multiples of these.)

10. $(-2, 1, 5)$ 11. $(11, 13, -7)$

12. (a) $X = (1, 0, -1) + t(-2, 1, 5)$
    (b) $X = (-10, -13, 7) + t(11, 13, -7)$ or also $(1, 0, 0) + t(11, 13, -7)$

13. (a) $-\frac{1}{3}$ (b) $-\dfrac{2}{\sqrt{42}}$ (c) $\dfrac{4}{\sqrt{66}}$ (d) $-\dfrac{2}{\sqrt{18}}$

14. (a) $(-4, \frac{11}{2}, \frac{15}{2})$ (b) $(\frac{25}{13}, \frac{10}{13}, -\frac{9}{13})$ 15. $(1, 3, -2)$

16. (a) $\dfrac{8}{\sqrt{35}}$ (b) $\dfrac{13}{\sqrt{21}}$

II, §1, p. 46

1. $A + B = \begin{pmatrix} 0 & 7 & 1 \\ 0 & 1 & 1 \end{pmatrix}$, $3B = \begin{pmatrix} -3 & 15 & -6 \\ 3 & 3 & -3 \end{pmatrix}$

$-2B = \begin{pmatrix} 2 & -10 & 4 \\ -2 & -2 & 2 \end{pmatrix}$, $A + 2B = \begin{pmatrix} -1 & 12 & -1 \\ 1 & 2 & 0 \end{pmatrix}$

$2A + B = \begin{pmatrix} 1 & 9 & 4 \\ -1 & 1 & 3 \end{pmatrix}$, $A - B = \begin{pmatrix} 2 & -3 & 5 \\ -2 & -1 & 3 \end{pmatrix}$

$A - 2B = \begin{pmatrix} 3 & -8 & 7 \\ -3 & -2 & 4 \end{pmatrix}$, $B - A = \begin{pmatrix} -2 & 3 & -5 \\ 2 & 1 & -3 \end{pmatrix}$

2. $A + B = \begin{pmatrix} 0 & 0 \\ 2 & -2 \end{pmatrix}$, $3B = \begin{pmatrix} -3 & 3 \\ 0 & -9 \end{pmatrix}$, $-2B = \begin{pmatrix} 2 & -2 \\ 0 & 6 \end{pmatrix}$,

$A + 2B = \begin{pmatrix} -1 & 1 \\ 2 & -5 \end{pmatrix}$, $A - B = \begin{pmatrix} 2 & -2 \\ 2 & 4 \end{pmatrix}$, $B - A = \begin{pmatrix} -2 & 2 \\ -2 & -4 \end{pmatrix}$

Rows of A: $(1, 2, 3)$, $(-1, 0, 2)$

Columns of A: $\begin{pmatrix} 1 \\ -1 \end{pmatrix}, \begin{pmatrix} 2 \\ 0 \end{pmatrix}, \begin{pmatrix} 3 \\ 2 \end{pmatrix}$

Rows of B: $(-1, 5, -2)$, $(1, 1, -1)$

Columns of B: $\begin{pmatrix} -1 \\ 1 \end{pmatrix}$, $\begin{pmatrix} 5 \\ 1 \end{pmatrix}$, $\begin{pmatrix} -2 \\ -1 \end{pmatrix}$

Rows of A: $(1, -1)$, $(2, 1)$ Columns of A: $\begin{pmatrix} 1 \\ 2 \end{pmatrix}$, $\begin{pmatrix} -1 \\ 1 \end{pmatrix}$

Rows of B: $(-1, 1)$, $(0, -3)$ Columns of B: $\begin{pmatrix} -1 \\ 0 \end{pmatrix}$, $\begin{pmatrix} 1 \\ -3 \end{pmatrix}$

4. (a) ${}^{t}A = \begin{pmatrix} 1 & -1 \\ 2 & 0 \\ 3 & 2 \end{pmatrix}$, ${}^{t}B = \begin{pmatrix} -1 & 1 \\ 5 & 1 \\ -2 & -1 \end{pmatrix}$

(b) ${}^{t}A = \begin{pmatrix} 1 & 2 \\ -1 & 1 \end{pmatrix}$, ${}^{t}B = \begin{pmatrix} -1 & 0 \\ 1 & -3 \end{pmatrix}$

5. Let $c_{ij} = a_{ij} + b_{ij}$. The ij-component of ${}^{t}(A + B)$ is $c_{ji} = a_{ji} + b_{ji}$, which is the sum of the ji-component of A plus the ji-component of B.

7. Same   8. $\begin{pmatrix} 0 & 2 \\ 0 & -2 \end{pmatrix}$, same

9. $A + {}^{t}A = \begin{pmatrix} 2 & 1 \\ 1 & 2 \end{pmatrix}$, $B + {}^{t}B = \begin{pmatrix} -2 & 1 \\ 1 & -6 \end{pmatrix}$

10. (a) ${}^{t}(A + {}^{t}A) = {}^{t}A + {}^{tt}A = {}^{t}A + A = A + {}^{t}A$.
(b) ${}^{t}(A - {}^{t}A) = {}^{t}A - {}^{tt}A = -(A - {}^{t}A)$
(c) The diagonal elements are 0 because they satisfy

$$a_{ii} = -a_{ii}.$$

II, §2, p. 58

1. $IA = AI = A$ 2. O

3. (a) $\begin{pmatrix} 3 & 2 \\ 4 & 1 \end{pmatrix}$ (b) $\begin{pmatrix} 10 \\ 14 \end{pmatrix}$ (c) $\begin{pmatrix} 33 & 37 \\ 11 & -18 \end{pmatrix}$

5. $AB = \begin{pmatrix} 4 & 2 \\ 5 & -1 \end{pmatrix}$, $BA = \begin{pmatrix} 2 & 4 \\ 4 & 1 \end{pmatrix}$

6. $AC = CA = \begin{pmatrix} 7 & 14 \\ 21 & -7 \end{pmatrix}$, $BC = CB = \begin{pmatrix} 14 & 0 \\ 7 & 7 \end{pmatrix}$

If $C = xI$, where x is a number, then $AC = CA = xA$.

7. $(3, 1, 5)$, first row

8. Second row, third row, i-th row

9. (a) $\begin{pmatrix} 2 \\ 4 \end{pmatrix}$ (b) $\begin{pmatrix} 4 \\ 6 \end{pmatrix}$ (c) $\begin{pmatrix} 3 \\ 5 \end{pmatrix}$

10. (a) $\begin{pmatrix} 3 \\ 1 \\ 2 \end{pmatrix}$ (b) $\begin{pmatrix} 12 \\ 3 \\ 9 \end{pmatrix}$ (c) $\begin{pmatrix} 5 \\ 4 \\ 8 \end{pmatrix}$

11. Second column of A **12.** j-th column of A

13. (a) $\begin{pmatrix} 4 \\ 9 \\ 5 \end{pmatrix}$ (b) $\begin{pmatrix} 3 \\ 1 \end{pmatrix}$ (c) $\begin{pmatrix} x_2 \\ 0 \end{pmatrix}$ (d) $\begin{pmatrix} 0 \\ x_1 \end{pmatrix}$

14. (a) $\begin{pmatrix} a & ax+b \\ c & cx+d \end{pmatrix}$. Add a multiple of the first column to the second column.
Other cases are similar.

16. (a) $A^2 = \begin{pmatrix} 0 & 0 & 1 \\ 0 & 0 & 0 \\ 0 & 0 & 0 \end{pmatrix}$, $A^3 = O$ matrix. If $B = \begin{pmatrix} 0 & 1 & 1 & 1 \\ 0 & 0 & 1 & 1 \\ 0 & 0 & 0 & 1 \\ 0 & 0 & 0 & 0 \end{pmatrix}$ then

$B^2 = \begin{pmatrix} 0 & 0 & 1 & 2 \\ 0 & 0 & 0 & 1 \\ 0 & 0 & 0 & 0 \\ 0 & 0 & 0 & 0 \end{pmatrix}$, $B^3 = \begin{pmatrix} 0 & 0 & 0 & 1 \\ 0 & 0 & 0 & 0 \\ 0 & 0 & 0 & 0 \\ 0 & 0 & 0 & 0 \end{pmatrix}$ and $B^4 = O$.

(b) $A^2 = \begin{pmatrix} 1 & 2 & 3 \\ 0 & 1 & 2 \\ 0 & 0 & 1 \end{pmatrix}$, $A^3 = \begin{pmatrix} 1 & 3 & 6 \\ 0 & 1 & 3 \\ 0 & 0 & 1 \end{pmatrix}$, $A^4 = \begin{pmatrix} 1 & 4 & 10 \\ 0 & 1 & 4 \\ 0 & 0 & 1 \end{pmatrix}$

17. $\begin{pmatrix} 1 & 0 & 0 \\ 0 & 4 & 0 \\ 0 & 0 & 9 \end{pmatrix}$, $\begin{pmatrix} 1 & 0 & 0 \\ 0 & 8 & 0 \\ 0 & 0 & 27 \end{pmatrix}$, $\begin{pmatrix} 1 & 0 & 0 \\ 0 & 16 & 0 \\ 0 & 0 & 81 \end{pmatrix}$

18. Diagonal matrix with diagonal $a_1^k, a_2^k, \ldots, a_n^k$.

19. 0, 0

20. (a) $\begin{pmatrix} 0 & 1 \\ -1 & 0 \end{pmatrix}$

(b) $\begin{pmatrix} a & b \\ -a^2/b & -a \end{pmatrix}$ for any $a, b \neq 0$; if $b = 0$, then $\begin{pmatrix} 0 & 0 \\ c & 0 \end{pmatrix}$.

21. (a) Inverse is $I + A$.
(b) Multiply $I - A$ by $I + A + A^2$ on each side. What do you get?

22. (a) Multiply each side of the relation $B = TAT^{-1}$ on the left by T^{-1} and on the right by T. We get

$$T^{-1}BT = T^{-1}TAT^{-1}T = IAI = A.$$

Hence there exists a matrix, namely T^{-1}, such that $T^{-1}BT = A$. This means that B is similar to A.

(b) Suppose A has the inverse A^{-1}. Then $TA^{-1}T^{-1}$ is an inverse for B because

$$TA^{-1}T^{-1}B = TA^{-1}T^{-1}TAT^{-1} = TA^{-1}AT^{-1} = TT^{-1} = I.$$

And similarly $BTA^{-1}T^{-1} = I$.

(c) Take the transpose of the relation $B = TAT^{-1}$. We get

$${}^tB = {}^tT^{-1}\,{}^tA\,{}^tT.$$

This means that ${}^tB$ is similar to ${}^tA$, because there exists a matrix, namely ${}^tT^{-1} = C$, such that ${}^tB = CAC^{-1}$.

23. Diagonal elements are $a_{11}b_{11},\ldots,a_{nn}b_{nn}$. They multiply componentwise.

24. $\begin{pmatrix} 1 & a+b \\ 0 & 1 \end{pmatrix}, \begin{pmatrix} 1 & na \\ 0 & 1 \end{pmatrix}$

25. $\begin{pmatrix} 1 & -a \\ 0 & 1 \end{pmatrix}$

26. Multiply AB on each side by $B^{-1}A^{-1}$. What do you get? Note the order in which the inverses are taken.

27. (a) The addition formula for cosine is

$$\cos(\theta_1 + \theta_2) = \cos\theta_1\cos\theta_2 - \sin\theta_1\sin\theta_2.$$

This and the formula for the sine will give what you want.

(b) $A(\theta)^{-1} = A(-\theta)$. Multiply $A(\theta)$ by $A(-\theta)$, what do you get?

(c) $A^n = \begin{pmatrix} \cos n\theta & -\sin n\theta \\ \sin n\theta & \cos n\theta \end{pmatrix}$. You can prove this by induction. Take the product of A^n with A. What do you get?

28. (a) $\begin{pmatrix} 0 & -1 \\ 1 & 0 \end{pmatrix}$ (b) $\dfrac{1}{\sqrt{2}}\begin{pmatrix} 1 & -1 \\ 1 & 1 \end{pmatrix}$ (c) $\begin{pmatrix} -1 & 0 \\ 0 & -1 \end{pmatrix}$ (d) $\begin{pmatrix} -1 & 0 \\ 0 & -1 \end{pmatrix}$

(e) $\dfrac{1}{2}\begin{pmatrix} 1 & \sqrt{3} \\ -\sqrt{3} & 1 \end{pmatrix}$ (f) $\dfrac{1}{2}\begin{pmatrix} \sqrt{3} & -1 \\ 1 & \sqrt{3} \end{pmatrix}$ (g) $\dfrac{1}{\sqrt{2}}\begin{pmatrix} -1 & 1 \\ -1 & -1 \end{pmatrix}$

29. $\begin{pmatrix} \cos\theta & \sin\theta \\ -\sin\theta & \cos\theta \end{pmatrix}$ **30.** $\dfrac{1}{\sqrt{2}}(-1, 3)$ **31.** $(-3, -1)$

32. The coordinates of Y are given by

$$y_1 = x_1\cos\theta - x_2\sin\theta,$$
$$y_2 = x_1\sin\theta + x_2\cos\theta.$$

Find $y_1^2 + y_2^2$ by expanding out, using simple arithmetic. Lots of terms will cancel out to leave $x_1^2 + x_2^2$.

33. (a) $\begin{pmatrix} 1 & 4 & 2 & -2 \\ 0 & 0 & 0 & 0 \\ 0 & 0 & 0 & 0 \\ 0 & 0 & 0 & 0 \end{pmatrix}$ (b) $\begin{pmatrix} 0 & 0 & 0 & 0 \\ 2 & 3 & -1 & 1 \\ 0 & 0 & 0 & 0 \\ 0 & 0 & 0 & 0 \end{pmatrix}$

34. (a) Interchange first and second row of A.
 (b) Interchange second and third row of A.
 (c) Add five times second row to fourth row of A.
 (d) Add -2 times second row to third row of A.

35. (a) Multiply first row of A by 3.
 (b) Add 3 times third row to first row.
 (c) Subtract 2 times first row from second row.
 (d) Subtract 2 times second row from third row.

36. (a) Put s-th row of A in r-th place, zeros elsewhere.
 (b) Interchange r-th and s-th rows, put zeros elsewhere.
 (c) Interchange r-th and s-th rows.

37. (a) Add 3 times s-th row to r-th row.
 (b) Add c times s-th row to r-th row.

II, §3, p. 69

1. Let $X = (x_1,\ldots,x_n)$. Then $X \cdot E_i = x_i$, so if this 0 for all i then $x_i = 0$ for all i.

3. $X \cdot (c_1 A_1 + \cdots + c_n A_n) = c_1 X \cdot A_1 + \cdots + c_n X \cdot A_n = 0$.

II, §4, p. 76

(There are several possible answers to the row echelon form, we give one of them. Others are also correct.)

1. (a) $\begin{pmatrix} 1 & 2 & -5 \\ 0 & 9 & -26 \\ 0 & 0 & 0 \end{pmatrix}$ and also $\begin{pmatrix} 1 & 0 & \frac{7}{9} \\ 0 & 1 & -\frac{26}{9} \\ 0 & 0 & 0 \end{pmatrix}$

 (b) $\begin{pmatrix} 1 & 0 & 2 \\ 0 & -1 & -1 \\ 0 & 0 & -1 \end{pmatrix}$ and also $\begin{pmatrix} 1 & 0 & 0 \\ 0 & 1 & 0 \\ 0 & 0 & 1 \end{pmatrix}$

2. (a) $\begin{pmatrix} 1 & -2 & 3 & -1 \\ 0 & 3 & -4 & 4 \\ 0 & 0 & 7 & -10 \end{pmatrix}$ and also $\begin{pmatrix} 1 & 0 & 0 & \frac{15}{7} \\ 0 & 1 & 0 & -\frac{4}{7} \\ 0 & 0 & 1 & -\frac{10}{7} \end{pmatrix}$

 (b) $\begin{pmatrix} 2 & 0 & -7 & 5 \\ 0 & 1 & 3 & -2 \\ 0 & 0 & 0 & 0 \end{pmatrix}$ and also $\begin{pmatrix} 1 & 0 & -\frac{7}{2} & \frac{5}{2} \\ 0 & 1 & 3 & -2 \\ 0 & 0 & 0 & 0 \end{pmatrix}$

3. (a) $\begin{pmatrix} 1 & 2 & -1 & 2 & 1 \\ 0 & 0 & 3 & -6 & 1 \\ 0 & 0 & 0 & -6 & 1 \end{pmatrix}$ or also $\begin{pmatrix} 1 & 2 & 0 & 0 & \frac{4}{3} \\ 0 & 0 & 1 & 0 & 0 \\ 0 & 0 & 0 & 1 & -\frac{1}{6} \end{pmatrix}$

(b) $\begin{pmatrix} 1 & 3 & -1 & 2 \\ 0 & 11 & -5 & 3 \\ 0 & 0 & 0 & 0 \\ 0 & 0 & 0 & 0 \end{pmatrix}$ or also $\begin{pmatrix} 1 & 0 & \frac{4}{11} & \frac{13}{11} \\ 0 & 1 & -\frac{5}{11} & \frac{3}{11} \\ 0 & 0 & 0 & 0 \\ 0 & 0 & 0 & 0 \end{pmatrix}$

II, §5, p. 85

1. (a) $-\dfrac{1}{20}\begin{pmatrix} 4 & 1 & -7 \\ -4 & -6 & 2 \\ -12 & 2 & 6 \end{pmatrix}$ **(b)** $\dfrac{1}{5}\begin{pmatrix} 2 & 23 & -11 \\ 1 & 19 & -8 \\ 0 & -10 & 5 \end{pmatrix}$

(c) $\dfrac{1}{4}\begin{pmatrix} 3 & 2 & -9 \\ 1 & 2 & -3 \\ -2 & -4 & 10 \end{pmatrix}$ **(d)** $\dfrac{1}{5}\begin{pmatrix} 5 & -16 & 3 \\ 0 & 7 & -1 \\ 0 & -2 & 1 \end{pmatrix}$

(e) $-\dfrac{1}{76}\begin{pmatrix} 0 & -19 & 0 \\ -32 & -14 & 12 \\ 28 & 17 & -20 \end{pmatrix}$

(f) $-\dfrac{1}{14}\begin{pmatrix} -17 & 7 & -9 \\ 11 & -7 & 5 \\ 13 & -7 & 11 \end{pmatrix}$

2. The effect of multiplication by I_{rs} is to put the s-th row in the r-th place, and zeros elsewhere. Thus the s-th row of $I_{rs}A$ is 0. Multiplying by I_{rs} once more puts 0 in the r-th row, and 0 elsewhere, so $I_{rs}^2 = O$.

3. We have $E_{rs}(c) = I + cI_{rs}$ and $E_{rs}(c') = I + c'I_{rs}$ so

$$E_{rs}(c)E_{rs}(c') = (I + cI_{rs})(I + c'I_{rs})$$
$$= I + cI_{rs} + c'I_{rs} + cc'I_{rs}^2$$
$$= I + (c + c')I_{rs} \qquad \text{because} \qquad I_{rs}^2 = O$$
$$= E_{rs}(c + c').$$

III, §1, p. 93

1. Let B and C be perpendicular to A_i for all i. Then

$$(B + C)\cdot A_i = B\cdot A_i + C\cdot A_i = 0 \qquad \text{for all } i.$$

Also for any number x,

$$(xB)\cdot A_i = x(B\cdot A_i) = 0.$$

Finally $O\cdot A_i = O$ for all i. This proves that W is a subspace.

2. (c) Let W be the set of all (x, y) such that $x + 4y = 0$. Elements of W are then of the form $(-4y, y)$. Letting $y = 0$ shows that $(0, 0)$ is in W. If $(-4y, y)$ and $(-4y', y')$ are in W, then their sum is $(-4(y + y'), y + y')$ and so lies in W. If c is a number, then $c(-4y, y) = (-4cy, cy)$, which lies in W. Hence W is a subspace.

4. Let v_1, v_2 be in the intersection $U \cap W$. Then their sum $v_1 + v_2$ is both in U (because v_1, v_2 are in U) and in W (because v_1, v_2 are in W) so is in the intersection $U \cap W$. We leave the other conditions to you.

Now let us prove partially that $U + W$ is a subspace. Let u_1, u_2 be elements of U and w_1, w_2 be elements of W. Then

$$(u_1 + w_1) + (u_2 + w_2) = u_1 + u_2 + w_1 + w_2,$$

and this has the form $u + w$, with $u = u_1 + u_2$ in U and $w = w_1 + w_2$ in W. So the sum of two elements in $U + W$ is also in $U + W$. We leave the other conditions to the reader.

5. Let A and B be perpendicular to all elements of V. Let X be an element of V. Then $(A + B) \cdot X = A \cdot X + B \cdot X = 0$, so $A + B$ is perpendicular to all elements of V. Let c be a number. Then $(cA) \cdot X = c(A \cdot X) = 0$, so cA is perpendicular to all elements of V. This proves that the set of elements of $\mathbf{R}^n$ perpendicular to all elements of V is a subspace.

III, §4, p. 109

2. (a) $A - B$, $(1, -1)$ (b) $\frac{1}{2}A + \frac{3}{2}B$, $(\frac{1}{2}, \frac{3}{2})$
   (c) $A + B$, $(1, 1)$ (d) $3A + 2B$, $(3, 2)$
3. (a) $(\frac{1}{3}, -\frac{1}{3}, \frac{1}{3})$ (b) $(1, 0, 1)$ (c) $(\frac{1}{3}, -\frac{1}{3}, -\frac{2}{3})$
4. Assume that $ad - bc \neq 0$. Let $A = (a, b)$ and $C = (c, d)$. Suppose we have

$$xA + yC = O.$$

This means in terms of coordinates

$$xa + yc = 0,$$
$$xb + yd = 0.$$

Multiply the first equation by d, the second by c and subtract. We find

$$x(ad - bc) = 0.$$

Since $ad - bc \neq 0$ this implies that $x = 0$. A similar elimination shows that $y = 0$. This proves (i).

Conversely, suppose A, C are linearly independent. Then neither of them can be $(0, 0)$ (otherwise pick x, $y \neq 0$, and get $xA + yC = 0$ which is impossible). Say b or $d \neq 0$. Then

$$d(a, b) - b(c, d) = (ad - bc, 0).$$

Since A, C are assumed linearly independent, the right-hand side cannot be 0, so $ad - bc \neq 0$. The argument is similar if a or $c \neq 0$.

For (iii), given an arbitrary vector (s, t), solve the system of linear equations arising from $xA + yC = (s, t)$ by elimination. You will find precisely that you need $ad - bc \neq 0$ to do so.

6. Look at Chapter I, §4, Exercise 7.

9. $(3, 5)$

10. $(-5, 3)$

11. Possible basis: $\begin{pmatrix} 1 & 0 \\ 0 & 0 \end{pmatrix}, \begin{pmatrix} 0 & 1 \\ 0 & 0 \end{pmatrix}, \begin{pmatrix} 0 & 0 \\ 1 & 0 \end{pmatrix}, \begin{pmatrix} 0 & 0 \\ 0 & 1 \end{pmatrix}$

12. $\{E_{ij}\}$ where E_{ij} has component 1 at the (i, j) place and 0 otherwise. These elements generate $\text{Mat}(m \times n)$, because given any matrix $A = (a_{ij})$ we can write it as a linear combination

$$A = \sum_i \sum_j a_{ij} E_{ij}.$$

Furthermore, if

$$O = \sum_i \sum_j a_{ij} E_{ij}$$

then we must have $a_{ij} = 0$ for all indices i, j so the elements E_{ij} are linearly independent.

13. E_i where E_i is the $n \times n$ matrix whose ii-th term is 1 and all other terms are 0.

14. A basis can be chosen to consist of the elements E_{ij} having ij-component equal to 1 for $i \leq j$ and all other components equal to 0. The number of such elements is

$$1 + 2 + \cdots + n = \frac{n(n + 1)}{2}.$$

15. (a) $\begin{pmatrix} 0 & 1 \\ 1 & 0 \end{pmatrix}, \begin{pmatrix} 1 & 0 \\ 0 & 0 \end{pmatrix}, \begin{pmatrix} 0 & 0 \\ 0 & 1 \end{pmatrix}$

(b) $\left\{ \begin{pmatrix} 1 & 0 & 0 \\ 0 & 0 & 0 \\ 0 & 0 & 0 \end{pmatrix} \begin{pmatrix} 0 & 0 & 0 \\ 0 & 1 & 0 \\ 0 & 0 & 0 \end{pmatrix} \begin{pmatrix} 0 & 0 & 0 \\ 0 & 0 & 0 \\ 0 & 0 & 1 \end{pmatrix} \begin{pmatrix} 0 & 1 & 0 \\ 1 & 0 & 0 \\ 0 & 0 & 0 \end{pmatrix} \begin{pmatrix} 0 & 0 & 1 \\ 0 & 0 & 0 \\ 1 & 0 & 0 \end{pmatrix} \begin{pmatrix} 0 & 0 & 0 \\ 0 & 0 & 1 \\ 0 & 1 & 0 \end{pmatrix} \right\}$

16. A basis for the space $\text{Sym}(n \times n)$ of symmetric $n \times n$ matrices can be taken to be the elements E_{ij} with $i \leq j$ having ij-component equal to 1, ji-component equal to 1, and rs-component equal to 0 if $(r, s) \neq (i, j)$ or (j, i). The proof that these generate $\text{Sym}(n \times n)$ and are linearly independent is similar to the proof in Exercise 12.

III, §5, p. 115

1. (a) 4   (b) mn (c) n (d) $n(n + 1)/2$ (e) 3   (f) 6   (g) $n(n + 1)/2$

2. 0, 1, or 2, by Theorem 5.8. The subspace consists of O alone if and only if it has dimension 0. If the subspace has dimension 1, let v_1 be a basis. Then the subspace consists of all elements tv_1, for all numbers t, so is a line by definition. If the subspace has dimension 2, let v_1, v_2 be a basis. Then the subspace consists of all elements $t_1v_1 + t_2v_2$ where t_1, t_2 are numbers, so is a plane by definition.

3. 0, 1, 2, or 3 by Theorem 5.8.

III, §6, p. 121

1. (a) 2   (b) 2   (c) 2   (d) 1   (e) 2   (f) 3   (g) 3   (h) 2   (i) 2

IV, §1, p. 126

1. (a) $\cos x$ (b) e^x (c) $1/x$ 2. $(-1/\sqrt{2}, 1/\sqrt{2})$

3. (a) 11   (b) 13   (c) 6

4. (a) $(e, 1)$ (b) $(1, 0)$ (c) $(1/e, -1)$

5. (a) $(e + 1, 3)$ (b) $(e^2 + 2, 6)$ (c) $(1, 0)$

6. (a) $(2, 0)$ (b) $(\pi e, \pi)$

7. (a) 1   (b) 11

8. Ellipse $9x^2 + 4y^2 = 36$ 9. Line $x = 2y$

10. Circle $x^2 + y^2 = e^2$, circle $x^2 + y^2 = e^{2c}$

11. Cylinder, radius 1, z-axis = axis of cylinder   12. Circle $x^2 + y^2 = 1$

IV, §2, p. 134

1. All except (c), (g)

2. Only Exercise 8.

5. Since $AX = BX$ for all X this relation is true in particular when $X = E^j$ is the j-th unit vector. But then $AE^j = A^j$ is the j-th column of A, and $BE^j = B^j$ is the j-th column of B, so $A^j = B^j$ for all j. This proves that $A = B$.

6. Only $u = O$, because $T_u(O) = u$ and if T_u is linear, then we must have $T_u(O) = O$.

7. The line S can be represented in the form $P + tv_1$ with all numbers t. Then $L(S)$ consists of all points

$$L(P) + tL(v_1).$$

If $L(v_1) = O$, this is a single point. If $L(v_1) \neq O$, this is a line. Other cases done similarly.

8. Parallelogram whose vertices are B, $3A$, $3A + B$, O.

9. Parallelogram whose vertices are 0, $2B$, $5A$, $5A + 2B$.

10. (a) $(-1, -1)$ (b) $(-2/3, 1)$ (c) $(-2, -1)$

11. (a) $(4, 5)$ (b) $(11/3, -3)$ (c) $(4, 2)$

12. Suppose we have a relation $\sum x_i v_i = O$. Apply F. We obtain $\sum x_i F(v_i) = \sum x_i w_i = O$. Since the w_i's are linearly independent it follows that all $x_i = 0$.

13. (a) Let v be an arbitrary element of V. Since $F(v_0) \neq 0$ there exists a number c such that

$$F(v) = cF(v_0),$$

namely $c = F(v)/F(v_0)$. Then $F(v - cv_0) = 0$, so let $w = v - cv_0$. We have written $v = w + cv_0$ as desired.
   (b) W is a subspace by Exercise 3. By part (a), the elements $v_0, v_1, \ldots, v_n$ generate V. Suppose there is a linear relation

$$c_0 v_0 + c_1 v_1 + \cdots + c_n v_n = O.$$

Apply F. We get $c_0 F(v_0) = 0$. Since $F(v_0) \neq 0$ it follows that $c_0 = 0$. But then $c_i = 0$ for $i = 1, \ldots, n$ because $v_1, \ldots, v_n$ form a basis of W.

IV, §3, p. 141

1 and 2. If U is a subspace of V then dim $L(U) \leq$ dim U. Hence the image of a one-dimensional subspace is either 0 or 1. The image of a two-dimensional subspace is 0, 1, or 2. A line or plane is of the form $P + U$, where U has dimension 1 or 2. Its image is of the form $L(P) + L(U)$, so the assertions are now clear.

3. (a) By the dimension formula, the image of F has dimension n. By Theorem 4.6 of Chapter III, the image must be all of W.
   (b) is similar.

4. Use the dimension formula.

5. Since $L(v_0 + u) = L(v_0)$ if u is in Ker L, every element of the form $v_0 + u$ is a solution. Conversely, let v be a solution of $L(v) = w$. Then

$$L(v - v_0) = L(v) - L(v_0) = w - w = O,$$

so $v - v_0 = u$ is in the kernel, and $v = v_0 + u$.

6. Constant functions.

7. Ker D^2 = polynomials of deg ≤ 1, Ker D^n = polynomials of deg $\leq n - 1$.

8. (a) Constant multiples of e^x (b) Constant multiples of e^{ax}

9. (a) $n - 1$ (b) $n^2 - 1$

10. $A = \dfrac{A + {}^tA}{2} + \dfrac{A - {}^tA}{2}$. If $A = B + C = B_1 + C_1$, then

$$B - B_1 = C_1 - C.$$

But $B - B_1 = C_1 - C$ is both symmetric and skew-symmetric, so O because each component is equal to its own negative.

11. (c) Taking the transpose of $(A + {}^tA)/2$ show that this is a symmetric matrix. Conversely, given a symmetric matrix B, we see that $B = P(B)$, so B is in the image of P.
  (d) $n(n - 1)/2$.
  (e) A basis for the skew-symmetric matrices consists of the matrices E_{ij} with $i < j$ having ij-component equal to 1, ji-component equal to -1, and all other components equal to 0.

12. Similar to 11.

13 and **14.** Similar to 11 and 12.

15. (a) 0   (b) $m + n$, $\{(u_i, 0), (0, w_j)\}$; $i = 1,\dots,m$; $j = 1,\dots,n$. If $\{u_i\}$ is a basis of U and $\{w_j\}$ is a basis of W.

16. (b) The image is clearly contained in $U + W$. Given an arbitrary element $u + w$ with u in U and w in W, we can write it in the form $u + w = u - (-w)$, which shows that it is in the image of L.
  (c) The kernel of L consists of those elements (u, w) such that $u - w = O$, so $u = w$. In other words, it consists of the pairs (u, u), and u must lie both in U and W, so in the intersection. If $\{u_1,\dots,u_r\}$ is a basis for $U \cap W$, then $\{(u_1, u_1),\dots,(u_r, u_r)\}$ is a basis for the kernel of L. The dimension is the same as the dimension of $U \cap W$. Then apply the dimension formula in the text.

IV, §4, p. 149

1. $n - 1$ **2.** 4   **3.** $n - 1$

4. (a) dim. = 1   basis = $(1, -1, 0)$
  (b) dim. = 2   basis = $(1, 1, 0)(0, 1, 1)$
  (c) dim. = 1   basis = $\left(\dfrac{\pi - 3}{10}, \dfrac{\pi + 2}{5}, 1\right)$
  (d) dim. = 0

5. (a) 1   (b) 1   (c) 0   (d) 2

6. One theorem states that

$$\dim V = \dim \text{Im } L + \dim \text{Ker } L.$$

Since $\dim \text{Ker } L \geqq 0$, the desired inequality follows.

7. One proof (there are others): rank $A = \dim \text{Im } L_A$. But $L_{AB} = L_A \circ L_B$. Hence the image of L_{AB} is contained in the image of L_A. Hence rank $AB \leqq$ rank A.

For the other inequality, note that the rank of a matrix is equal to the rank of its transpose, because column rank equals row rank. Hence

$$\text{rank } AB = \text{rank } {}^tB^tA.$$

Now apply the first inequality to get rank ${}^tB^tA \leqq$ rank ${}^tB =$ rank B.

IV, §5, p. 156

1. (a) $\begin{pmatrix} 1 & 0 & 0 & 0 \\ 0 & 1 & 0 & 0 \end{pmatrix}$ (b) $\begin{pmatrix} 1 & 0 & 0 & 0 \\ 0 & 1 & 0 & 0 \\ 0 & 0 & 1 & 0 \end{pmatrix}$

(c) $3I$ (d) $7I$ (e) $-I$ (f) $\begin{pmatrix} 1 & 0 & 0 & 0 \\ 0 & 1 & 0 & 0 \\ 0 & 0 & 0 & 0 \\ 0 & 0 & 0 & 0 \end{pmatrix}$

2. cI, where I is the unit $n \times n$ matrix.

3. (a) $\begin{pmatrix} 1 & -4 & 3 \\ -3 & 2 & 1 \end{pmatrix}$ (b) $\begin{pmatrix} 3 & -2 & 1 \\ 4 & -1 & 5 \end{pmatrix}$

4. (a) $\begin{pmatrix} 0 & 1 & -2 \\ 3 & -2 & 4 \\ -1 & 1 & 5 \end{pmatrix}$ (b) $\begin{pmatrix} 3 & 0 & 0 \\ 0 & -7 & 0 \\ 0 & 0 & 5 \end{pmatrix}$ (c) $\begin{pmatrix} -2 & 0 & 1 \\ 0 & 0 & 0 \\ 7 & -1 & 0 \end{pmatrix}$

5. Let $Lv_i = \sum c_{ij}w_j$. Let $C = (c_{ij})$. The associated matrix is ${}^tC$, and the effect of L on a coordinate vector X is ${}^tCX$.

6. $\begin{bmatrix} c_1 & 0 & \cdots & 0 \\ 0 & c_2 & \cdots & 0 \\ \vdots & \vdots & & \vdots \\ 0 & 0 & \cdots & c_n \end{bmatrix}$ **7.** $\begin{pmatrix} 0 & 1 \\ -1 & 0 \end{pmatrix}$ **8.** $\begin{bmatrix} 0 & 1 & 0 \\ 0 & 0 & 2 \\ 0 & 0 & 0 \end{bmatrix}$

V, §1, p. 162

1. Let $C = A - B$. Then $CX = O$ for all X. Take $X = E^j$ to be the j-th unit vector for $j = 1,\ldots,n$. Then $CE^j = C^j$ is the j-th column of C. By assumption, $CE^j = O$ for all j so $C = O$.

2. Use distributivity and the fact that $F \circ L = L \circ F$.

3. Same proof as with numbers.

4. $P^2 = \frac{1}{4}(I + T)^2 = \frac{1}{4}(I^2 + 2TI + T^2) = \frac{1}{4}(2I + 2T) = \frac{1}{2}(I + T) = P$. Part (b) is left to you. For part (c), see the next problem.

5. (a) $Q^2 = (I - P)^2 = I^2 - 2IP + P^2 = I - 2P + P = I - P = Q$.
   (b) Let $v \in \operatorname{Im} P$ so $v = Pw$ for some w. Then $Qv = QPv = 0$ because $QP = (I - P)P = P - P^2 = P - P = 0$. Hence $\operatorname{Im} P \subset \operatorname{Ker} Q$. Conversely, let $v \in \operatorname{Ker} Q$ so $Qv = 0$. Then $(I - P)v = 0$ so $v - Pv = 0$, and $v = Pv$, so $v \in \operatorname{Im} P$, whence $\operatorname{Ker} Q \subset \operatorname{Im} P$.

6. Let $v \in V$. Then $v = v - Pv + Pv$, and $v - Pv \in \operatorname{Ker} P$ because

$$P(v - Pv) = Pv - P^2v = Pv - Pv = 0.$$

Furthermore $Pv \in \operatorname{Im} P$, thus proving (a).

As to (b), let $v \in \operatorname{Im} P \cap \operatorname{Ker} P$. Since $v \in \operatorname{Im} P$ there exists $w \in V$ such that $v = Pw$. Since $v \in \operatorname{Ker} P$, we get

$$0 = Pv = P^2w = Pw = v,$$

so $v = 0$, whence (b) is also proved.

7. Suppose $u + w = u_1 + w_1$. Then $u - u_1 = w_1 - w$. But $u - u_1 \in U$ and $w_1 - w \in W$ because U, W are subspaces. By assumption that $U \cap W = \{0\}$, we conclude that $u - u_1 = 0 = w_1 - w$ so $u = u_1$, $w = w_1$.

8. $P^2(u, w) = P(u, 0) = (u, 0) = P(u, w)$. So $P^2 = P$.

9. The dimension of a subspace is $\leq$ the dimension of the space. Then

$$\operatorname{Im} F \circ L \leq \operatorname{Im} F \quad \text{so} \quad \dim \operatorname{Im} F \circ L \leq \dim \operatorname{Im} F$$

so rank $F \circ L \leq$ rank F. This proves one of the formulas. For the other, view F as a linear map defined on $\operatorname{Im} L$. Then

$$\operatorname{rank} F \circ L = \dim \operatorname{Im} F \circ L \leq \dim \operatorname{Im} L = \operatorname{rank} L.$$

V, §2, p. 168

1. $R_\theta^{-1} = R_{-\theta}$ because $R_\theta \circ R_{-\theta} = R_{\theta - \theta} = R_0 = I$. The matrix associated with R_θ^{-1} is

$$\begin{pmatrix} \cos\theta & \sin\theta \\ -\sin\theta & \cos\theta \end{pmatrix}$$

because $\cos(-\theta) = \cos\theta$.

3. The composites as follows are the identity:

$$F \circ G \circ G^{-1} \circ F^{-1} = I \quad \text{and} \quad G^{-1} \circ F^{-1} \circ F \circ G = I.$$

4, 5, 6. In each case show that the kernel is O and apply the appropriate theorem.

7 through 10. The proof is similar to the same proof for matrices, using distributivity. In **7**, we have

$$(I - L) \circ (I + L) = I^2 - L^2 = I.$$

For **8**, we have $L^2 + 2L = -I$ so $L(-L - 2I) = I$, so $L^{-1} = -L - 2I$.

11. It suffices to prove that v, w are linearly independent. Suppose $xv + yw = 0$. Apply L. Then

$$L(w) = L(L(v)) = O$$

because $L^2 = O$. Hence $L(xv) = xL(v) = O$. Since $L(v) \neq O$, it follows that $x = 0$. Then $y = 0$ because $w \neq O$.

12. F is injective, its kernel is 0.
(b) F is not surjective, for instance $(1, 0, 0, \ldots)$ is not in the image.
(c) Let $G(x_1, x_2, \ldots) = (x_2, x_3, \ldots)$ (drop the first coordinate).
(d) No, otherwise F would have an inverse, which it does not.

13. Linearity is easily checked. To show that L is injective, it suffices to show that $\operatorname{Ker} L = \{0\}$. Suppose $L(u, w) = 0$, then $u + w = 0$, so $u = -w$. By assumption $U \cap W = \{0\}$, and $u \in U$, $-w \in W$ so $u = w = 0$. Hence $\operatorname{Ker} L = \{0\}$.
L is surjective because by assumption, every element of V can be written as the sum of an element of U and an element of W.

VI, §1, p. 178

2. Let $X = {}^t(x, y)$. Then

$$\langle X, X \rangle = ax^2 + 2bxy + dy^2 = a\left(x + \frac{b}{a} y\right)^2 + \frac{(ad - b^2)}{a} y^2.$$

If $ad - b^2 > 0$, then $\langle X, X \rangle$ is a sum of squares with positive coefficients, and one of the two terms is not 0, so $\langle X, X \rangle > 0$. If $ad - b^2 \leqq 0$, then give y any non-zero value, and let $x = -by/a$. Then $\langle X, X \rangle \leqq 0$.
If $a = 0$, let $y = 0$ and give x any value. Then ${}^tXAX = 0$.

3. (a) yes (b) no (c) yes (d) no (e) yes (f) yes

4. (a) 2 (b) 4 (c) 8

5. The diagonal elements are unchanged under the transpose, so the trace of A and ${}^tA$ is the same.

6. Let $A A = (c_{ij})$. Then

$$c_{ii} = \sum_{k=1}^{n} a_{ik} a_{ki}.$$

Hence

$$\mathrm{tr}(AA) = \sum_{i=1}^{n} \sum_{k=1}^{n} a_{ik} a_{ki},$$

and $a_{ik} a_{ki} = a_{ik}^2$ because A is symmetric, so the trace is a sum of squares, hence ≥ 0.

8. (a)

$$\mathrm{tr}(AB) = \sum_{i} \sum_{j} a_{ij} b_{ji} = \sum_{j} \sum_{i} b_{ji} a_{ij}.$$

But any pair of letters can be used to denote indices in this sum, which can be rewritten more neutrally.

$$\sum_{r=1}^{n} \sum_{s=1}^{n} b_{rs} a_{sr}.$$

This is precisely the trace $\mathrm{tr}(BA)$.
(b) $\mathrm{tr}(C^{-1}AC) = \mathrm{tr}(ACC^{-1})$ by part (a), so $= \mathrm{tr}(A)$.

VI, §2, p. 189

1. (a) $\dfrac{1}{\sqrt{3}}$ (1, 1, −1) and $\dfrac{1}{\sqrt{2}}$ (1, 0, 1)

(b) $\dfrac{1}{\sqrt{6}}$ (2, 1, 1), $\dfrac{1}{5\sqrt{3}}$ (−1, 7, −5)

2. $\dfrac{1}{\sqrt{6}}$ (1, 2, 1, 0) and $\dfrac{1}{\sqrt{39}}$ (−1, −2, 5, 3)

3. $\dfrac{1}{\sqrt{2}}$ (1, 1, 0, 0), $\tfrac{1}{2}$(1, −1, 1, 1), $\dfrac{1}{\sqrt{18}}$ (−2, 2, 3, 1)

5. $\sqrt{80}$ $(t^2 - 3t/4)$, $\sqrt{3}\, t$

6. $\sqrt{80}$ $(t^2 - 3t/4)$, $\sqrt{3}\, t$, $10t^2 - 12t + 3$

9. Use the dimension formulas. The trace is a linear map, from V to **R**. Since

$$\dim V = \dim \mathrm{Ker}\ \mathrm{tr} + \dim \mathrm{Im}\ \mathrm{tr},$$

it follows that $\dim W = \dim V - 1$ so $\dim W^{\perp} = 1$ by Theorem 2.3. Let I be the unit $n \times n$ matrix. Then $\mathrm{tr}(A) = \mathrm{tr}(AI)$ for all $A \in V$, so $\mathrm{tr}(AI) = 0$ for $A \in W$. Hence $I \in W^{\perp}$. Since $W^{\perp}$ has dimension 1, it follows that I is a basis of $W^{\perp}$. (Simple?)

10. We have $\langle X, AY \rangle = \langle X, bY \rangle = b \langle X, Y \rangle$ and also

$$\langle X, AY \rangle = \langle AX, Y \rangle = \langle aX, Y \rangle = a \langle X, Y \rangle.$$

So $a \langle X, Y \rangle = b \langle X, Y \rangle$. Since $a \neq b$ it follows that $\langle X, Y \rangle = 0$.

VI, §3, p. 194

1. For all column vectors X, Y we note that ${}^tXAY$ is a 1×1 matrix, so equal to its own transpose. Therefore

$$\varphi_A(X, Y) = {}^tXAY = {}^t({}^tXAY) = {}^tY^tA^{tt}X = {}^tYAX = \varphi_A(Y, X).$$

2. Conversely, if $\varphi_A(X, Y) = \varphi_A(Y, X)$, then a similar argument as above shows that

$$^tXAY = {}^tX^tAY$$

for all X, Y. Then $A = {}^tA$ by the proof of uniqueness in Theorem 3.1.

3. (a) $2x_1y_1 - 3x_1y_2 + 4x_2y_1 + x_2y_2$
   (b) $4x_1y_1 + x_1y_2 - 2x_2y_1 + 5x_2y_2$
   (c) $5x_1y_1 + 2x_1y_2 + \pi x_2y_1 + 7x_2y_2$
   (d) $x_1y_1 + 2x_1y_2 - x_1y_3 - 3x_2y_1 + x_2y_2 + 4x_2y_3 + 2x_3y_1 + 5x_3y_2 - x_3y_3$
   (e) $-4x_1y_1 + 2x_1y_2 + x_1y_3 + 3x_2y_1 + x_2y_2 + x_2y_3 + + 2x_3y_1 + 5x_3y_2 + 7x_3y_3$
   (f) $-\frac{1}{2}x_1y_1 + 2x_1y_2 - 5x_1y_3 + x_2y_1 + \frac{2}{3}x_2y_2 + 4x_2y_3 - x_3y_1 + 3x_3y_3$

VII, §2, p. 207

1. (a) -20 (b) 5 (c) 4 (d) 16 (e) -76 (f) -14

2. (a) -15 (b) 45 (c) 0 (d) 0 (e) 4 (f) 14 (g) 108 (h) 135 (i) 10

3. $a_{11}a_{22}\cdots a_{nn}$ 4. 1

5. Even in the 3×3 case, follow the general directives. Subtract the second column times x_1 from the third. Subtract the first column times x_1 from the second. You get

$$\begin{vmatrix} 1 & 0 & 0 \\ 1 & x_2 - x_1 & x_2^2 - x_2x_1 \\ 1 & x_3 - x_1 & x_3^2 - x_3x_1 \end{vmatrix}.$$

Expanding according to the first row yields

$$\begin{vmatrix} x_2 - x_1 & x_2(x_2 - x_1) \\ x_3 - x_1 & x_3(x_3 - x_1) \end{vmatrix}.$$

You can factor out $x_2 - x_1$ from the first row and $x_3 - x_1$ from the second row to get

$$(x_2 - x_1)(x_3 - x_1)\begin{vmatrix} 1 & x_2 \\ 1 & x_3 \end{vmatrix}.$$

Then use the formula for 2×2 determinants to get the desired answer.
   Now do the 4×4 case in detail to understand the pattern. Then do the $n \times n$ case by induction.

6. (a) 3 (b) -24 (c) 16 (d) 14 (e) 0 (f) 8 (g) 40 (h) -10 (i) $\prod_{i=1}^{N} a_{ii}$

7. 1 **8.** $t^2 + 8t + 5$

11. $D(cA) = D(cA^1, cA^2, cA^3) = cD(A^1, cA^2, cA^3) = c^3 D(A^1, A^2, A^3)$ using the linearity with respect to each column.

12. $D(A) = D(cA^1, \ldots, cA^n) = c^n D(A^1, \ldots, A^n)$ using again the linearity with respect to each column.

VII, §3, p. 214

1. 2 **2.** 2 **3.** 2 **4.** 3 **5.** 4 **6.** 3 **7.** 2 **8.** 3

VII, §4, p. 217

1. (a) $x = -\frac{1}{3}$, $y = \frac{2}{3}$, $z = -\frac{1}{3}$ (b) $x = \frac{5}{12}$, $y = -\frac{1}{12}$, $z = \frac{1}{12}$
    (c) $x = -\frac{5}{24}$, $y = \frac{97}{48}$, $z = \frac{1}{3}$, $w = -\frac{25}{48}$
    (d) $x = \frac{11}{2}$, $y = \frac{38}{5}$, $z = \frac{1}{10}$, $w = 2$

VII, §5, p. 221

1. Chapter II, §5

2. $\dfrac{1}{ad - bc} \begin{pmatrix} d & -b \\ -c & a \end{pmatrix}$

VII, §6, p. 232

2. (a) 14 (b) 1

3. (a) 11 (b) 38 (c) 8 (d) 1

4. (a) 10 (b) 22 (c) 11 (d) 0

VIII, §1, p. 237

1. Let $\begin{pmatrix} x \\ y \end{pmatrix}$ be an eigenvector. Then matrix multiplication shows that we must have $x + ay = \lambda x$ and $y = \lambda y$. If $y \neq 0$ then $\lambda = 1$. Since $a \neq 0$ this contradicts the first equation, so $y = 0$. Then E^1, which is an eigenvector, forms a basis for the space of eigenvectors.

2. If $A = cI$ is a scalar multiple of the identity, then the whole space consists of eigenvectors. Any basis of the whole space answers the requirements. The only eigenvalue is c itself, for non-zero vectors.

3. The unit vector E^i is an eigenvector, with eigenvalue a_{ii}. The set of these unit vectors is a basis for the whole space.

4. Let $y = 1$. Then using matrix multiplication, you find that $'(x, 1)$ is an eigenvector, with eigenvalue 1. If $\theta = 0$, the unit vectors E^1, E^2 are eigenvectors, with eigenvalues 1 and -1 respectively.

5. Let $v_2 = {}'(-1, x)$ so v_2 is perpendicular to v_1 in Exercise 4. Matrix multiplication shows that $Av_2 = -v_2$.

6. Let $A = \begin{pmatrix} a & -b \\ b & a \end{pmatrix}$. Then the characteristic polynomial is $(t - a)^2 + b^2$, and for it to be equal to 0 we must have $b = 0$, $t = a$. But $a^2 + b^2 = 1$ so $a = \pm 1$.

7. $A(Bv) = ABv = BAv = B\lambda v = \lambda Bv$.

VIII, §2, p. 249

1. (a) $(t - a_{11})\ldots(t - a_{nn})$ (b) $a_{11},\ldots,a_{nn}$

2. Same as in Exercise 1.

3. (a) $(t - 4)(t + 1)$; eigenvalues 4, -1; corresponding eigenvectors $\begin{pmatrix} 2 \\ 3 \end{pmatrix}$ and $\begin{pmatrix} 1 \\ -1 \end{pmatrix}$, or non-zero scalar multiples.

(b) $(t - 1)(t - 2)$; eigenvalues 1, 2; eigenvectors $(1, -1/\lambda)$ with $\lambda = 1$ and $\lambda = 2$.

(c) $t^2 + 3$; eigenvalues $\pm\sqrt{-3}$; eigenvectors $(1, 1/(\lambda - 2))$ with $\lambda = \sqrt{-3}$ and $= -\sqrt{-3}$.

(d) $(t - 5)(t + 1)$; eigenvalues 5, -1; eigenvectors $\begin{pmatrix} 1 \\ 1 \end{pmatrix}$ and $\begin{pmatrix} 2 \\ -1 \end{pmatrix}$ respectively, or non-zero scalar multiples.

4. (a) $(t - 1)(t - 2)(t - 3)$; eigenvalues 1, 2, 3; eigenvectors

$$\begin{pmatrix} 0 \\ 1 \\ 0 \end{pmatrix}, \quad \begin{pmatrix} -1/2 \\ 1 \\ 1 \end{pmatrix}, \quad \begin{pmatrix} -1 \\ 1 \\ 1 \end{pmatrix}$$

respectively, or non-zero scalar multiples.

(b) $(t - 4)(t + 2)^2$; eigenvalues 4, -2; Eigenvectors:

$$\begin{pmatrix} 1 \\ 1 \\ 2 \end{pmatrix} \text{ for 4;} \quad \begin{pmatrix} 1 \\ 1 \\ 0 \end{pmatrix} \text{ and } \begin{pmatrix} 1 \\ 0 \\ -1 \end{pmatrix} \text{ for } -2.$$

Non-zero scalar multiples of the first; linear combinations of the pair of eigenvectors for -2 are also possible, or in general, the space of solutions of the equation

$$x - y + z = 0.$$

(c) $(t - 2)^2(t - 6)$; eigenvalues 2, 6; eigenvectors:

$$\begin{pmatrix} 1 \\ -1 \\ 0 \end{pmatrix} \quad \text{and} \quad \begin{pmatrix} 1 \\ 0 \\ -1 \end{pmatrix} \text{ for 2;} \quad \begin{pmatrix} 1 \\ 2 \\ 1 \end{pmatrix} \text{ for 6.}$$

Linear combinations of the first two are possible. Non-zero scalar multiples of the second are possible.

(d) $(t - 1)(t - 3)^2$; eigenvalues 1, 3; eigenvectors:

$$\begin{pmatrix} 2 \\ -1 \\ 1 \end{pmatrix} \text{ for 1;} \quad \begin{pmatrix} 1 \\ 1 \\ 0 \end{pmatrix} \quad \text{and} \quad \begin{pmatrix} 1 \\ 0 \\ 1 \end{pmatrix} \text{ for 3.}$$

Non-zero scalar multiples of the first are possible. Linear combinations of the second two are possible.

5. (a) Eigenvalue 1; eigenvectors scalar multiples of $'(1, 1)$.
   (b) Eigenvalue 1; eigenvectors scalar multiples of $'(1, 0)$.
   (c) Eigenvalue 2; eigenvectors scalar multiples of $'(0, 2)$.
   (d) Eigenvalues $\lambda_1 = (1 + \sqrt{-3})/2$, $\lambda_2 = (1 - \sqrt{-3})/2$; Eigenvectors scalar multiples of $'(1, (\lambda + 1)^{-1})$ with $\lambda = \lambda_1$ or $\lambda = \lambda_2$. There is no real eigenvector.

6. Eigenvalues 1 in all cases. Eigenvectors scalar multiples of $'(1, 0, 0)$.

7. (a) Eigenvalues ± 1, $\pm\sqrt{-1}$. Eigenvectors scalar multiples of $'(1, \lambda, \lambda^2, \lambda^3)$ where λ is any one of the four eigenvalues. There are only two real eigenvalues.
   (b) Eigenvalues 2, $(-1 + \sqrt{-3})/2$, $(-1 - \sqrt{-3})/2$. Eigenvectors scalar multiples of $\,{}^t\!\left(1, \dfrac{(\lambda + 1)^2 + 4}{13}, \lambda + 1\right)$ where λ is one of the three eigenvalues. There is only one real eigenvector namely $'(1, 1, 3)$ up to real scalar multiples.

9. Each root of the characteristic polynomial is an eigenvalue. Hence A has n distinct eigenvalues by assumption. By Theorem 1.2 n corresponding eigenvectors are linearly independent. Since dim $V = n$ by assumption, these eigenvectors must be a basis.

10. We assume from the chapter on determinants that the determinant of a matrix is equal to the determinant of its transpose. Then

$$\det(xI - A) = \det(xI - {}^tA),$$

so the roots of the characteristic polynomial of A are the same as the roots of the characteristic polynomial of ${}^tA$.

11. That λ is an eigenvalue of A means there exists a vector $v \neq 0$ such that $Av = \lambda v$. Since A is assumed invertible, the kernel of A is 0, so $\lambda \neq 0$. Apply A^{-1} to this last equation. We get

$$v = A^{-1}(\lambda v) = \lambda A^{-1} v,$$

whence $A^{-1} v = \lambda^{-1} v$, so λ^{-1} is an eigenvalue of A^{-1}.

12. Let $f(t) = \sin t$ and $g(t) = \cos t$. Then

$$Df = 0 + g,$$
$$Dg = -f + 0.$$

Hence the matrix associated with D with respect to the basis $\{f, g\}$ is

$$A = \begin{pmatrix} 0 & -1 \\ 1 & 0 \end{pmatrix}.$$

The characteristic polynomial is

$$P_A(t) = \begin{vmatrix} t & 1 \\ -1 & t \end{vmatrix} = t^2 + 1.$$

Since this polynomial has no root in $\mathbf{R}$, it follows that A, and hence D, has no eigenvalue in the 2-dimension space whose basis is $\{f, g\}$.

13. By calculus, $d^2(\sin kx)/dx^2 = -k^2 \sin kx$. This means that the function $f(x) = \sin kx$ is an eigenvector of D^2, with eigenvalue $-k^2$. Similarly for the function $g(x) = \cos kx$.

14. Let v be an eigenvector of A, so that $Av = \lambda v$, $v \neq 0$. Since $\{v_1, \dots, v_n\}$ is a basis of V, there exist numbers $a_1, \dots, a_n$ such that

$$v = a_1 v_1 + \cdots + a_n v_n.$$

Applying A yields

$$Av = \lambda v = a_1 c_1 v_1 + \cdots + a_n c_n v_n.$$

But we also have

$$\lambda v = \lambda a_1 v_1 + \cdots + \lambda a_n v_n.$$

Subtracting gives

$$a_1(\lambda - c_1)v_1 + \cdots + a_n(\lambda - c_n)v_n = 0.$$

Since $v_1, \ldots, v_n$ are linearly independent, we must have

$$a_i(\lambda - c_i) = 0 \qquad \text{for all} \quad i = 1, \ldots, n.$$

Say $a_1 \neq 0$. Then $\lambda - c_1 = 0$ so $\lambda = c_1$. Since we assumed that the numbers $c_1, \ldots, c_n$ are distinct, it follows that $\lambda - c_j \neq 0$ for $j = 2, \ldots, n$ whence $a_j = 0$ for $j = 2, \ldots, n$. Hence finally $v = a_1 v_1$. This concludes the proof.

15. Let $v \neq 0$ be an eigenvector for AB with eigenvalue λ, so that $ABv = \lambda v$. Then

$$BABv = \lambda Bv.$$

If $Bv \neq 0$ then Bv is an eigenvector for BA with this same eigenvalue λ. If on the other hand $Bv = 0$, then $\lambda = 0$. Furthermore, BA cannot be invertible (otherwise if C is an inverse, $BAC = I$ so B is invertible, which is not the case). Hence there is a vector $w \neq 0$ such that $BAw = 0$, so 0 is also an eigenvalue of BA. This proves that the eigenvalues of AB are also eigenvalues of BA.

There is an even better proof of a more general fact, namely:

The characteristic polynomial of AB is the same as that of BA.

To prove this, suppose first that B is invertible. Since the determinant of a product is the product of the determinants, we get:

$$\det(xI - AB) = \det(B(xI - AB)B^{-1}) = \det(xI - BABB^{-1}) = \det(xI - BA).$$

This proves the theorem assuming B invertible. But the identity to be proved

$$\det(xI - AB) = \det(xI - BA)$$

is a polynomial identity, in which we can consider the components of A and B as "variables", so if the identity is true when A is fixed and $B = (b_{ij})$ is "variable" then it is true for all B. A matrix with "variable" components is invertible, so the previous argument applies to conclude the proof.

VIII, §3, p. 255

1. (a) 1, 3

   (b) $(1 + \sqrt{5})/2$, $(1 - \sqrt{5})/2$

2. (a) 0, 1, 3

   (b) 2, $2 \pm \sqrt{2}$

   The maximum value of f is the largest number in each case.

3. $\dfrac{-1 \pm \sqrt{74}}{2}$

VIII, §4, p. 259

1. $\sum a_i x_i^2$ if $a_1, \ldots, a_n$ are the diagonal elements.

2. Let B have diagonal elements $\lambda_1^{1/2}, \ldots, \lambda_n^{1/2}$.

3. Let v be an eigenvector $\neq O$ with eigenvalue λ. Then $\langle Av, v \rangle = \langle \lambda v, v \rangle = \langle v, v \rangle$. Since $\langle v, v \rangle > 0$ and $\langle Av, v \rangle > 0$, it follows that $\lambda > 0$. Pick a basis of V consisting of eigenvectors. The vector space V can then be identified as the space of coordinate vectors with respect to this basis. The matrix of A then is a diagonal matrix, whose diagonal elements are the eigenvalues, and are therefore positive. We can then use Exercise 2 to find a square root.

4. Similar to Exercise 3.

5. From ${}^t(AA) = {}^tA{}^tA = AA$, it follows that A^2 is symmetric. Furthermore, for $v \neq O$,

$$\langle A^2 v, v \rangle = \langle Av, {}^tAv \rangle = \langle Av, Av \rangle > 0,$$

because $Av \neq O$ since $\langle Av, v \rangle > O$.

Since ${}^tA^{-1} = A^{-1}$ it follows that A^{-1} is symmetric. Since A is invertible, a given v can be written $v = Aw$ for some w (namely $w = A^{-1}v$). Then

$$\langle A^{-1}v, v \rangle = \langle A^{-1}Aw, Aw \rangle = \langle w, Aw \rangle = \langle Aw, w \rangle > 0.$$

Hence A^{-1} is positive definite.

6. Assume (i). From the identity in the hint, we get

$$4\langle Uv, Uw \rangle = \langle U(v + w), U(v + w) \rangle - \langle U(v - w), U(v - w) \rangle$$
$$= \langle (v + w), v + w) \rangle - \langle (v - w), (v - w) \rangle$$
$$= 4\langle v, w \rangle.$$

Hence $\langle Uv, Uw \rangle = \langle v, w \rangle$. The converse is immediate.

Assume (i). Then Ker $U = O$ because if $Uv = O$ then $\|v\| = 0$ so $v = O$. But a linear map with O kernel from a finite dimensional vector space into itself is invertible, so U is invertible. Also, for all $v, w \in V$,

$$\langle Uv, Uw \rangle = \langle {}^tUUv, w \rangle \qquad \text{and also equals } \langle v, w \rangle \text{ by hypothesis.}$$

Hence ${}^tUU = I$ so ${}^tU = U^{-1}$. Conversely, from ${}^tU = U^{-1}$ we get

$$\langle Uv, Uv \rangle = \langle {}^tUUv, v \rangle = \langle v, v \rangle,$$

so U satisfies (i).

7. Since $^t(^tAA) = {}^tA^{tt}A = {}^tAA$, it follows that ^tAA is symmetric. Furthermore, for $v \neq O$, we have

$$\langle {}^tAAv,\ v \rangle = \langle Av,\ Av \rangle > 0$$

because A is invertible, $Av \neq O$. Hence ^tAA is positive definite. Let $U = AB^{-1}$ where $B^2 = {}^tAA$ and $BA = AB$, so $B^{-1}A = AB^{-1}$. Then

$$\langle Uv,\ Uv \rangle = \langle AB^{-1}v,\ AB^{-1}v \rangle = \langle B^{-1}Av,\ B^{-1}Av \rangle$$
$$= \langle Av,\ {}^tB^{-1}B^{-1}Av \rangle$$
$$= \langle v,\ AA^{-2}Av \rangle = \langle v, v \rangle.$$

Hence U is unitary.

8. Let v be a non-zero eigenvalue $\lambda > 0$. Then

$$\langle Bv,\ Bv \rangle = \lambda^2 \langle v,\ v \rangle = \langle v,\ v \rangle$$

because B is unitary. Hence $\lambda^2 = 1$. Hence $\lambda = \pm 1$, and since λ is positive, $\lambda = 1$. Since V has a basis consisting of eigenvectors, it follows that $B = I$.

Index

A

Addition formula for sine and
   cosine 57
Additive inverse 45
d'Alembert 3
Alternating 204
Angle between planes 36
Angle between vectors 24
Area 222
Augmented matrix 71

B

Ball 19
Basis 105, 154, 180, 182
Beginning point 9
Bessel inequality 177
Bilinear 190
Box 98

C

Change of basis 154
Characteristic polynomial 238, 248
Characteristic value 233
Characteristic vector 233
Closed ball 19
Coefficient 85, 94
Column 43
Column operation 71
Column rank 115
Column space 115

Complex 243, 260
Complex conjugate 261
Component 24, 42, 174
Composition 158
Conjugate 261
Convex 99, 135
Coordinates 2, 107, 131
Cosine of angle 24
Cramer's rule 215

D

Determinant 195
Determinant and inverse 217
Determinant of linear map 230
Diagonal matrix 47
Diagonalization 255
Diderot 3
Differential equations 138, 258
Dimension 3, 111, 139, 144, 173, 187
Dimension of set of solutions 143, 173
Direct product 143
Direct sum 163
Direction 8, 11, 21
Distance 17
Distance between point and plane 39

E

Eigenfunction 136
Eigenspace 234
Eigenvalue 136, 233, 252

Eigenvector   136, 233
Element   88
Elementary matrix   63, 77, 80
Elementary row operation   71
End point   9
Expansion of determinant   200, 203

F

Fourier coefficient   174
Fourth dimension   3

G

Generate   92, 93
Gram–Schmidt orthogonalization   182

H

Half plane   100
Hessian   194
Homogeneous linear equations   67
Hyperplane   36

I

Identity map   128
Image   113, 136
Imaginary   251
Injective   166
Intersection   92
Inverse   55, 80, 218, 261
Invertible   164, 168
Isomorphism   168

K

Kernel   136
Kernel and Image   139

L

Leading coefficient   74
Length   173
Lie in   88
Line   30, 95, 112
Line segment   96
Linear combination   85, 94
Linear equation   49, 86
Linear mapping   127
Linearly dependent   86, 104, 210
Linearly independent   86, 104
Located vector   9, 10

M

Magnitude   16
Mapping   123
Markov matrix   50
Matrix   43
Matrix of linear map   150
Maximal set of linearly independent
    vectors   113
Maximum   251
Multilinear   204
Multiplication of matrices   47

N

Non-trivial combination   85
Non-trivial solution   67
Norm   15, 173
Normal to plane   35

O

Open ball   19
Opposite direction   8, 11
Orthogonal   14, 172, 180
Orthogonal complement   146
Orthonormal   180
Orthonormal basis   182

P

Parallel   10
Parallel planes   36
Parallelogram   6, 97, 173
Parametric representation   30
Perpendicular   12, 14, 34, 172
Perpendicular planes   36
Plane   34, 95, 112
Point in space   2
Position vector   30
Positive definite   171, 178, 259
Product   47
Product space   243
Projection   23, 129, 142, 163
Pythagoras theorem   22, 173

Q

Quadratic form   251

R

Rank   120, 145, 164, 212
Real part   261

Rotation 56, 159
Row 43
Row echelon form 73
Row equivalent 72
Row operation 70, 116
Row rank 115, 145
Row space 115

S

Same direction 8, 11
Scalar multiple 7
Scalar product 12, 171
Schwarz inequality 27, 175
Segment 30, 96
Set 88
Similar matrices 61, 156
Skew-symmetric matrix 47, 115, 142
Space 1
Spanned 102
Sphere 19
Square matrix 44
Standard unit vectors 21
Subset 88
Subspace 91
Sum of mappings 126
Sum of matrices 44
Sum of n-tuples 4
Surjective 166
Symmetric matrix 46, 115

T

Trace 178
Translation 95, 125, 159
Transpose 45, 151
Triangle inequality 28
Trivial solution 67

U

Unit matrix 53
Unit sphere 251
Unit vector 21, 173
Unknowns 67
Upper triangular matrix 62, 111, 115

V

Value 123
Vandermonde determinant 208
Vector 10, 89
Vector space 88
Volume 229

Z

Zero functions 90
Zero map 138
Zero matrix 44
Zero vector 5, 89

CPSIA information can be obtained
at www.ICGtesting.com
Printed in the USA
LVHW081138021218
598912LV00021B/58/P